GOOD FENGSHUI

A STEP-BY-STEP GUIDE TO CREATING BALANCE AND HARMONY IN YOUR HOME

EVA WONG

SHAMBHALA

SHAMBHALA PUBLICATIONS, INC.
2129 13th Street
Boulder, Colorado 80302
www.shambhala.com

Cover design: Katrina Noble
Interior design: Katrina Noble

9 8 7 6 5 4 3 2 1

First Edition
Printed in the United States of America

Shambhala Publications makes every effort to print on acid-free, recycled paper.
Shambhala Publications is distributed worldwide by Penguin Random House, Inc., and its subsidiaries.

LIBRARY OF CONGRESS CATALOGING-IN-PUBLICATION DATA
Names: Wong, Eva, 1951– author.
Title: Good fengshui: a step-by-step guide to creating balance and harmony in your home / Eva Wong.
Description: First edition. | Boulder, Colorado: Shambhala, 2023. | Includes index.
Identifiers: LCCN 2022036039 | ISBN 9781645470861 (trade paperback)
Subjects: LCSH: Feng shui in interior decoration.
Classification: LCC BF1779.F4 W6567 2023 | DDC 747—DC23/ENG/20220824
LC record available at https://lccn.loc.gov/2022036039

CONTENTS

PART THREE: Fengshui Strategies

PREFACE

IT'S BEEN A WHILE since I wrote my first two books on fengshui: *Feng-Shui: The Ancient Wisdom of Harmonious Wisdom for Modern Times* and *A Master Course in Feng-Shui*. At that time fengshui was not widely known outside Asian communities. Now, fengshui has become a household word and has made it into television talk shows and home improvement media. We are even seeing real estate contracts include statements like "pending fengshui approval."

My first book was designed to introduce readers to the historical and cultural background of fengshui and describe traditional fengshui techniques that are used to evaluate the energy of land forms, architecture, and floor plans. My goal was to help raise consciousness about fengshui, to show that the principles and practices of fengshui are based on the Chinese and especially the Taoist view that we live in a world of energy and that being in harmony with this energy can lead to good health and livelihood, an appreciation of the natural world, and a respect for others who share our environment with us.

My second book was written after I received numerous requests from the readership and my publisher to write a book teaching people how to do classical Chinese fengshui. The result was *A Master Course in Feng-Shui*—a book that is structured as a home-study course. The major focus of the book is a system of classical Chinese fengshui called Flying Stars (Xuan Kong). It was written during a time in which fengshui was beginning to attract those who wanted to learn a form of classical Chinese fengshui but did not have access to untranslated texts. Most of them proudly called themselves fengshui "nerds" and were drawn to a system of fengshui that is based on formulas and calculations. Flying Stars fengshui fits that demand perfectly.

We are now entering a different era of fengshui. It is starting to become a household word and is achieving a popularity that is beyond those who are academically interested and those who want to become professional fengshui consultants. There is a growing number of people who want to apply fengshui to building, purchasing, renting, and simply living in their home. Some are not interested in hiring consultants; others do not have local fengshui consultants to call on to help. While my home-study course teaches people to do fengshui for themselves, some feel that it is too technical to learn, or that they don't have the time to study the material thoroughly. People simply want to find a place with "good" fengshui and live there. They want advice that is practical and easy to follow. *Good Fengshui* is written exactly with that purpose in mind: to provide a practical form of fengshui that can be learned and applied by everyone.

There are a number of differences between this book and the previous ones I have written about fengshui. First, *Good Fengshui* focuses on aspects of fengshui that are practical and can be readily learned by everyone. The goal of this book is to present fengshui in a way that can be used by anyone who wants to find and live in a home with benevolent energy and avoid spaces with harmful energy. If you want handy tips on finding a good place to live or improving the space you are living in, I believe the information in this book will serve you well. If you cannot solve your fengshui problems with what you learn in this book, you can always work with a consultant. Chapter 17 provides advice on how to find a competent fengshui consultant.

Second, this book has an added focus on the rationale behind fengshui advice. Since writing my first two books, I have noticed a shift in how people approach fengshui. For a long time most people were content with recipes of what to do and what not to do. Now they want to know the "why" behind fengshui recommendations not just "how" to apply them. In this book, you'll find that in addition to recommendations, there are also discussions of the rationale behind them.

Lastly, having worked with many diverse clients in North America and Europe in the past twenty years, I have learned much about their fengshui needs. I would like to address their concerns, which are different from my Asian clients, especially those I consulted when I was living in Hong Kong. In this book you will find discussions of subjects that interest modern urban residents, such as the fengshui of neighborhoods, community life, influences of modern styles of architecture, and interior design.

Fengshui must be practical, and having been involved in the business of construction and engineering for the past ten years, I realize the importance of making fengshui recommendations that are practical and safe. Fengshui advice is only helpful when it can be implemented reasonably and safely. I do not give fengshui advice that is difficult to implement or unsafe structurally or architecturally.

This book is for anyone who wants to build, purchase, rent, or renovate a home or simply wants to "live" well in the space they're already in. Good fengshui means good health, good family relationships, good livelihood, and a good respect for the environment and those who live in it. Take this book with you when you hunt for a place to live. Look through it when you feel like doing a fengshui makeover of your space. Or simply enjoy reading something about fengshui that is entertaining and easy to understand.

I thank my clients and readers who have supported my fengshui business over the years. I have learned much from your suggestions, inquiries, and especially your commitment to making your home and your neighborhood a great place to live.

PART ONE

Talking about Fengshui

Fengshui was the subject of many lively conversations in the world I grew up in. Living in Hong Kong as a child, I was introduced to the word *fengshui* as early as I learned to read and write. My parents designed and arranged our home with the advice of a fengshui consultant—who happened to be my great uncle. This great uncle and his friend would become my fengshui teachers when I was nine.

Soon after I started my fengshui training, I discovered that fengshui was a hot topic in Hong Kong society. It was discussed in morning and late-night shows on television that were eagerly followed by over a million viewers. In crowded restaurants over morning tea and breakfast, you could hear snippets of conversations about the fengshui of the newest homes, commercial buildings, and even government offices. Over banquets, parties, and even on public transportation, you could hear people discussing good fengshui, bad fengshui, how fengshui can be fixed, and fengshui that is impossible to deal with.

Here in the United States, where I live now, homeowners, business entrepreneurs, architects, interior designers, and even real estate agents are beginning to use fengshui to build, design, and advise buyers and renters on how to choose a space that will bring health, wealth, and well-being. While fengshui is gaining popularity in non-Chinese societies, its origins and its cultural and philosophical backgrounds are still relatively unknown.

Before we discuss methods of how to choose a space with good fengshui, let's first talk about fengshui—how it originated, how it has evolved to what it is now, and why it is important to respect the environment we live in, appreciate the neighborhood and community that we are a part of, and make our own lives healthy, dignified, and uplifted.

1

What Is Fengshui?

FENGSHUI MEANS "wind and water." Gentle winds and clear water make an environment healthy, inviting, and prosperous. Healthy, because air is not stagnant and water is not polluted. Prosperous, because crops are bountiful, commerce is flourishing, and people are living together harmoniously.

In ancient China, towns and villages were built in places that had clean water and were sheltered from harsh winds. When these towns prospered, other people too began to want homes that were ventilated by gentle winds and were near clean water. No one wants to build a home where winds are violent, the air is stale, or waters are dirty and stagnant. This is just common sense. We all like gentle breezes and a view of peaceful clear water. No one wants to be buffeted by strong gusts of wind when they walk out of their home or be confronted by the smell of stinky stagnant water.

In Hong Kong, where I grew up, fengshui is a part of everyday life—but the popularity of fengshui is not just limited to Hong Kong (though it is the fengshui capital of the world). It is also prevalent in Taiwan, China, Singapore, and Chinese settlements in North America and Europe. Where there are Chinese people, there's fengshui. You might say that fengshui is a part of Chinese culture. For thousands of years the principles of fengshui have advised Chinese people on where to build towns, situate homes and businesses, and bury the dead. In each stage of life—birth, growth, death—fengshui is there as a trusted guide.

2

Where Did Fengshui Come From?

FENGSHUI HAS a long and illustrious history. Here are the highlights of major landmarks in the development of fengshui as a philosophy, a practice, and a culture.

Fengshui was practiced in ancient China as far back as four thousand years ago. Of course, it was not called "fengshui" then, but people used the same principles of fengshui that are used today to select places to settle and live. Based on common sense, experience, and trial and error, the Chinese found that settlements built in places where winds were gentle and water was clear were associated with good health, prosperity, and harmony.

With time, common sense and experiences became systematized into guidelines that were aligned with a culture that already saw land and all living things as carriers of energy, or qi. The study of qi in the land was called *kanyu* around three thousand years ago. *Kan* means "mountain" and *yu* means "valley." The study of kanyu blossomed into a sophisticated and elaborate system of identifying formations of mountains, valleys, and waterways that were associated with benevolent, malevolent, and neutral energy. The experts of this system of knowledge became known as geomancers, and it was their job to select sites for towns, villages, and even capital cities.

Around the time the emperor of China began to build the Great Wall, geomancers were tasked to find burial sites that would ensure the continuation of a dynasty and the prosperity of the nation. The rise of these imperial geomancers took kanyu away from the common people. Under the watchful eye of the government, kanyu could only be used by the emperor and the nobility. Commoners found using this knowledge were often punished by imprisonment or even death. The imperial families feared that a commoner might use the knowledge of kanyu to select a burial site for an ancestor that would start a new dynasty.

From the first to the tenth century, the practice of kanyu was used exclusively by imperial geomancers. However, in the eleventh and twelfth centuries, a social and cultural development in China changed the practice and outlook of kanyu forever. This was the rise of the merchant class and the expansion of urban settlements. Expansion of domestic and international trade routes, the emergence of a banking system, and the increase of employment prospects in cities brought on a population explosion in urban environments. More buildings had to be erected, and streets were soon crowded with homes and shops. As the merchants' wealth increased, so did their power. Even the nobility and the imperial government had to depend on them for taxes and levies for road construction, urban improvements, and even law enforcement.

The wealthy merchants began to entertain ideas on how to enhance their wealth and prosperity and create a mercantile dynasty for their descendants. Because they were not permitted to select burial sites for their ancestors, they began to consult with geomancers on how to build residential and business environments that would increase prosperity, enhance business prospects, and secure a future for the descendants. It is out of this social and cultural change that a new class of geomancers emerged: those who specialized in evaluating the energy of urban spaces.

Because urban environments are affected more by their immediate surroundings than by distant mountains and rivers, a new system of guidelines for building and designing homes and businesses emerged. This innovation was called *zhai hun*, meaning "the fortune of a building." Since, in a large city, you can hardly see natural waterways or even mountains from a home, as the only things generally visible are the homes and shops of your neighbors, the fortune of a particular building is more dependent on its immediate surroundings and the layout of the space within than on the geographical features in the land.

As China's commercial relationships expanded into other parts of Asia between the eleventh and thirteenth centuries, kanyu and zhai hun were introduced into the cultures of Vietnam, Korea, Japan, Tibet, and Thailand, influencing the siting and architecture of palaces, temples, homes, businesses, government buildings, and even urban design.

Meanwhile, in China, the popularity of zhai hun brought the use of geomancy back to the common people. By the sixteenth century, even the geomancy of burial ground selection was no longer restricted to the aristocratic classes. The discipline of kanyu became the study of land energy, and the techniques of zhai hun dominated building construction and urban design. By the eighteenth century, the knowledge and practice of these two domains of geomancy were called fengshui, based on the commonsense notion that an auspicious place to situate a building, whether for the living or the dead, was one that was nourished by gentle winds and clear water. Fengshui is no longer an exclusive commodity of the privileged classes. It is now for everyone.

And this is where we stand today. The knowledge and practice of constructing, selecting, and designing a building are inherited from several thousand years of experience on how to live a healthy, prosperous, and harmonious life.

Within the past fifty years, fengshui has spread from China across the world, as the Chinese migrated to Europe and the Americas. *Fengshui* is rapidly becoming a household word in Western culture, and more non-Chinese people are beginning to consult the principles of fengshui when building or selecting a home. In fact, the popularity of residential fengshui has grown so much in the past twenty years that it is common for property buyers and renters to hire fengshui consultants and for a condition of "fengshui approval" to appear in purchase and rental negotiations.

At this point, you might ask, If I want to evaluate the fengshui of a place before I buy or rent it, do I have to hire a consultant for every property I consider? Not necessarily. Most fengshui evaluations can be done by the home buyer or renter. In other words, *you* can do the fengshui yourself to a great extent. It's only when situations get complicated, or you are involved in large-scale building projects that you need to hire a professional fengshui consultant.

A typical buyer or renter can do enough fengshui themselves to ensure that what they end up with is beneficial and not harmful. This book is designed to help the renter and buyer to narrow down searches to decide where to live, as well as give advice to those who want to do a fengshui makeover of their home.

3

Commonsense Fengshui, Intuitive Fengshui, and Technical Fengshui

CAN ANYONE DO FENGSHUI? Yes. However, before you dive into it, you will want to know that there are three general approaches to fengshui. I call them commonsense fengshui, intuitive fengshui, and technical fengshui.

Commonsense fengshui is based on everyday intelligence and experience—we have a rationale for what makes a good or bad environment. Good environments allow us to move around comfortably and smoothly. Bad environments are obstructive and contribute to extra unnecessary activity.

This book is concerned with giving you guidelines for choosing and improving a home based on commonsense fengshui. My goal is to provide you with checklists of what to look for that constitutes good fengshui and what to avoid. Not all environments have perfect fengshui. Most have small issues that can be taken care of. In this book you will find advice on how to mitigate minor fengshui problems. There is no need to panic and move out if you find that your home does not have perfect fengshui.

Using the guidelines outlined in this book, you should be able to use commonsense fengshui competently to build, buy, or rent a home and even renovate your place. This is fengshui everyone can do.

Intuitive fengshui is based on gut feeling. We have a "feel" for what is a good environment and what is not. When we walk into a good environment, we feel comfortable, peaceful, and happy. On the other hand, when we walk into bad environments we feel nervous, uncomfortable, and even fearful. These feelings are instantaneous. There is no checklist, no rational deliberation, and not even second thoughts.

Intuitive fengshui is important because it allows us to tune our feelings in to aspects of the environment that cannot be explained rationally. We have all experienced feelings of uneasiness, anxiety, or fear when we enter certain spaces. Relying on gut feeling can help us avoid danger, obstacles, and problems. Intuitive fengshui can save us a lot of time and effort in eliminating buildings that have a feeling of "wrongness," although we can't tell what the "wrongness" is. I tell my clients, "If you don't feel right, don't move in." People have told me that they had walked into spaces where they felt goose bumps, chills, and even headaches. In comparison, others have told me that they have walked into spaces where they felt happy, peaceful, and safe.

You don't need a checklist for intuitive fengshui. You can cultivate your intuition by becoming more open to your feelings. Respect and trust your first reactions to a place. Most people don't trust their feelings because they have been told not to. Interestingly, children tend to have a better gut feeling about a place than adults. So, if you have children, bring them along when you look for a place to live. I am constantly amazed at how accurate children are in pinpointing fengshui problems.

Everyone can do intuitive fengshui. Just trust your instincts and feelings. First impressions and reactions are important. This is because those feelings are not filtered or censored by thoughts.

Technical fengshui is not for everyone. It is technical because it involves measurements, compass readings, and calculations based on terrestrial divination techniques developed in ancient China. This kind of fengshui is typically practiced by professional consultants and those who are inclined to study fengshui beyond the practical concerns of living comfortably and happily in an environment with good energy. The bottom line is this: if a place does not pass the test of commonsense and intuitive fengshui, there is no point in going into the technical details. Technical fengshui is useful after common sense and intuition have given you positive feedback about a space.

To further understand how energies from the direction a home faces and times in the calendar year affect your space, and learn how to enhance or mitigate problematic energies in each room, you can turn to technical fengshui. Flying Stars fengshui is the best technique for such in-depth analyses. My book *A Master Course in Feng-Shui* is a detailed study of how to apply technical fengshui to a home and a small business. If you are inclined toward technical fengshui, this home-study course is for you. Or, if you encounter problems you cannot solve with commonsense or intuitive fengshui, you can explore solutions using technical fengshui. Better yet, save yourself the effort and hire a professional consultant, but make sure the consultant is competent.

A list of guidelines on how to choose a professional consultant can be found in chapter 17.

4

Living with Fengshui Is Living with Energy

FENGSHUI WORKS because everything carries energy, or qi, and is connected to everything else. Land, human-made environments, architecture, plants, animals, humans, and even objects all carry some kind of energy. It is the interaction of these energies that makes a place beneficial, harmful, or neutral to live in.

Gentle, rolling land with greenery and vegetation is associated with beneficial energy while harsh, barren land with jagged rocks and steep slopes is associated with harmful energy. Flat land with little to no physical anomalies is associated with neutral energy that is neither beneficial nor harmful.

Human-made environments such as streets, parks, and mass transportation lines also carry energy. High-speed freeways, railway and subway lines, and runways are associated with fast, erratic, and restless energy. Low-speed meandering roads are associated with calm, smooth energy that is conducive to relaxed living. Maze-like streets are said to make people easily frustrated and irritable, and dark unlit spaces can introduce unwarranted fear and confusion. Parks, city squares, and open space allow energy to circulate freely. Narrow streets choke energy, and steep hills introduce frenetic and turbulent energies into an area.

Architecture also carries energy. Buildings with sharp-pointed features carry aggressive energy. Buildings with rounded characteristics carry beneficial energy. Buildings with nondescript features are said to have neutral energy. Architecture with strange and grotesque features is associated with fearful energy, and architecture with pleasant pleasing features is associated with harmonious and peaceful energy.

Plants and animals also carry energy. Leafy, blooming plants are associated with the energy of growth and nourishment. Plants with spines and aggressive-looking foliage are associated with unfriendly energy. Needless to say, it is desirable to live near fruit trees and not desirable to live near large cacti. It's not because certain plants are "bad." It's because the energy they carry is not friendly.

Friendly animals can bring nourishing, peaceful, and relaxing energy. I live in an area that is frequented by deer, rabbits, squirrels, and songbirds. Seeing them and hearing their calls bring me joy and peace. On the other hand, I have also stayed in jungles where the sound of snarling predators and the sight of wild boars have kept me restless and anxious, even though I was staying in a secure building with locks and gates.

Finally, there is human energy. Neighborhoods with aggressive and uncaring people make the environment feel threatening, while neighbors who are kind and helpful make everyone feel safe and secure. While we cannot change road patterns and the location of airports and transportation terminals, we can make our neighborhood peaceful and harmonious. If we can all be a little more sensitive, tolerant, cooperative, and helpful to others, it will make the environment we live in feel more safe and harmonious.

5

Environments Affect Us Physically and Mentally

SOME ENVIRONMENTS will support us, some will not do a thing for us, and some may even harm us. Some objects carry aggressive energy, others carry peaceful energy, and these energies can affect us physically and mentally. Because we and our environments are energetically linked, the influence of space on us is not just "psychological." No amount of psychological counseling can mitigate the harmful effects of bad fengshui. Some environments and objects do make us susceptible to illness, misfortunes, and mishaps, while others can bring us health, prosperity, harmony, and inner peace. These observations are not speculative; they are based on hundreds and even thousands of years of people's experiences. When many agree on the same thing over a long period of time, it is something real.

Fengshui is not just interior design. Interior design is about aesthetics; fengshui is about the energy a design carries. Some designs may be popular and trendy and may sell in the real estate market, but they could introduce negative energy into a space. Fengshui is not about what is trending; it is timeless.

The fact that you are reading this book means that you believe that fengshui can help you find a good place to live or improve the energy of the home you are currently living in. Good fengshui means good health, harmonious relationships, and good livelihood. Who doesn't want that?

6

Where Fengshui Shines

AS I MENTIONED, *fengshui* means "wind and water." Where there is gentle wind and flowing water there is nourishing qi, or life energy. Living in an environment with good qi in the land will enhance our health, livelihood, and relationships. However, where fengshui really shines is in the way we can use our knowledge of land energy to help ourselves and others appreciate the environment and understand the need to conserve it for future generations.

While land provides nourishment, neighborhoods build harmony and cooperation. Street patterns and the architecture of a neighborhood can influence how people relate to each other. Living in a neighborhood with architecture and road patterns that are associated with harmonious energy can facilitate cooperation and caring. Where fengshui shines here is in the way we can use the knowledge of neighborhood energy to help us appreciate the importance and the need to build harmonious communities.

Finally, our own homes are the source of health, well-being, and quality of life. Living in a home that is arranged to enhance positive energy and protect us from negative energy allows us to have an uplifted and relaxed lifestyle. Where fengshui shines in this case is in the way it can help us design a living space where we can look forward to waking up, resting well, having good meals, and enjoying the company of family and friends.

PART TWO

Fengshui You Can Do

Fengshui is about understanding energies and how to live with them. We want to embrace positive energy, avoid negative energy, and simply live with neutral energy.

Fengshui works with four kinds of energies. They are:

- Land energy
- People energy
- Architectural energy
- Object energy

These energies affect our health, livelihood, interpersonal relationships, and even our thoughts and emotions.

Land energy is energy carried by physical features in the land—mountains, hills, valleys, rivers, seas, lakes, and rock formations.

People energy is carried in cities and neighborhoods and is affected by neighborhood features such as street patterns, land use, and the types of human activities that dominate the area.

Architectural energy is manifested in the look and feel of the architecture of a building, its shape, its exterior design, and its interior layout or floor plan.

Object energy is carried by things, both natural and human made. In the external environment, objects, or things, include plants, decorative outdoor ornaments, and utility-related structures such as transformers, wires, and satellite dishes. Inside a building they include sculptures, art, fireplaces, decorative furniture, window and wall coverings, appliances, fans, and light fixtures.

Once you know how to identify positive, negative, and neutral energy in the land, in neighborhoods, in architecture, and in objects, you will be able to build, choose, renovate, or design a space that will give you the greatest benefits in health, prosperity, and interpersonal relationships.

7

Start with the Outside

ALWAYS START with the outside, the big picture. If the surrounding land and the neighborhood have problematic fengshui, then no matter how beautiful a house or an apartment looks, the negative energy carried in the environment will have an adverse impact on the building.

In this section you will learn how to evaluate the effects of land energy, people energy, architectural energy, and object energy. This is the fengshui of the environment. We evaluate the external environment first, because it has the most energetic impact on our lives.

Land Energy—The Fengshui of Land Formations

In evaluating land energy, we look at:

- Mountains and hills

- Valleys
- Natural waterways
- Human-made waterways
- Rock formations
- Animals and plant life

People Energy—The Fengshui of the Neighborhood

In evaluating people energy, we examine:

- Land use
- Types of neighborhoods
  - Mainly residential neighborhoods
  - Mainly commercial neighborhoods
  - Mixed residential and commercial neighborhoods
- Community spaces
  - Parks and recreation facilities
  - The town square
  - Markets
  - Public facilities
- Roads
  - Types of roads
  - Area road patterns
  - Local road patterns
  - The cul-de-sac
  - Train tracks, airport runways, bridges, tunnels, and elevated highways

Architectural Energy—The Fengshui of Buildings

In evaluating architectural energy, we examine:

- The architecture of the building
  - The look and feel of a building
  - The building materials
  - The shape of the building
- The types of residential buildings
  - Single homes
  - Duplex and row houses
  - Apartments
- The architecture of surrounding buildings
- The landscape features around the building

Object Energy—The Fengshui of Things

In evaluating object energy in the external environment, we look at:

- Outdoor landscape features
- Decorative outdoor ornaments
- Utility-related structures

8

Land Energy

THE FENGSHUI OF LAND FORMATIONS

LAND HAS THE STRONGEST influence on the fengshui of a place. Land affects our activities, feelings, and thoughts. Land should be your first consideration in selecting a place to live. You can change the color of your walls or even the floor plan of a home, but you can't change the surrounding landscape. So take some time to learn how to avoid land that does not benefit you and choose to live near land that will nourish you and enrich your lifestyle.

Here is a step-by-step guide on how to identify land that can enhance and land that can endanger your health, livelihood, personal development, and domestic and social life.

1. Look for interesting features in the land. What draws your eye? What arouses your curiosity and makes you want to explore it more? If you live around land that is interesting, you will be more creative, more curious about the world, and generally more inclined to try new things. If you live around land that does not have a variety of features, you will be less drawn to participating in new ventures and trying new alternatives. In a nutshell, if you live in a region with interesting land formations you will tend to be a more interesting person. You'll enjoy a richer domestic and social life. No one wants to follow someone on social media or invite them to parties if they are boring.
2. See if there is life on the land—trees, grasses, flowers, and wildlife. If the land does not nourish plant and animal life, it won't nourish you either.
3. To get a good view of the surrounding landscape, you will need to go to a vantage point where you can have a panoramic view.
4. The key to seeing how land affects the fengshui of a place is in identifying the kind of energy that is carried by different types of land formations.

Now let's evaluate the fengshui of land formations.

Mountains and Hills

Mountains and hills that are jagged and rocky carry sharp, aggressive, and destructive energy. These include knife-edge ridges, steep rocky slopes, and slabs of rock that protrude from the earth. These types of land can have an ambient negative energy for the neighborhood or even an entire town. Needless to say, it is not advisable to live adjacent to these kinds of formations.

Guidelines for Evaluating the Effect of Land Formations on a Building

- If you are totally surrounded by aggressive land formations, their impact will be the most severe.
- If these formations don't surround you but are close by, and you can see them directly, they will still affect you strongly.
- If you are farther away from the formations, but can still see them, their impact will be lessened but still noticeable.
- If you cannot see the formations at all, their impact will be minimal.

Land that is surrounded by gentle rolling hills without the presence of sharp jagged ridges carries gentle and nourishing energy. Therefore, it is desirable to live among or near this type of land formation.

Flat, uniform land carries dull energy. When everything around you is flat and featureless, it is boring. If the land is boring, your life will tend to be dull and boring as well.

Land consisting of fields of jumbled and crumbling rocks carries chaotic and restless energy. If you are

A. Land formations with negative energy

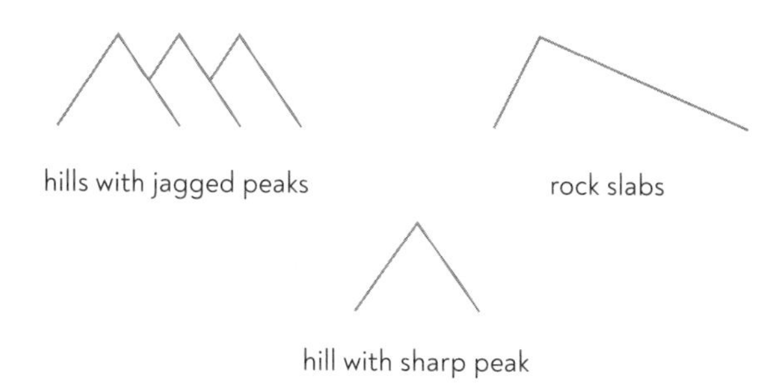

B. Land formations with nourishing energy

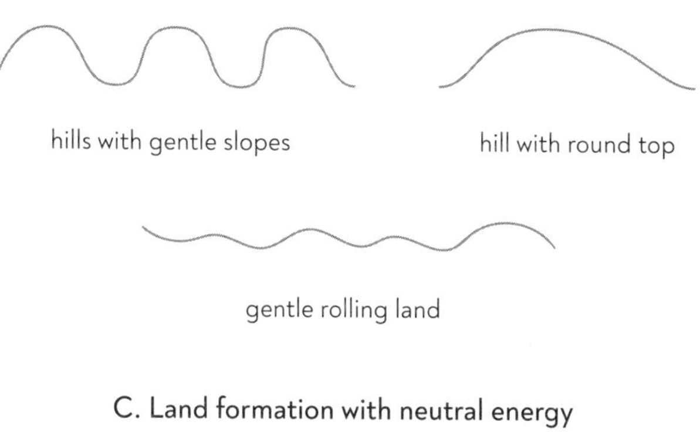

C. Land formation with neutral energy

flat land

FIGURE 1. Examples of mountain land forms with negative, nourishing, and neutral energy.

surrounded by chaotic energy, your lifestyle will tend to be restless, disorganized, and confused.

Valleys

Valleys include wide valleys, valleys with waterways, dry valleys, canyons, and even gullies.

Wide valleys with gentle slopes carry gentle and nourishing energy, especially if there are waterways running through them. Narrow valleys with steep slopes carry fast and restless energy.

Wide-bottomed valleys with steep slopes are called U-shaped valleys because they resemble the letter *U*. The slopes of the valley are cliff-like, forcing energy to flow only along the bottom of the valley. The steep and almost vertical walls prevent energy from spreading up the sides of the valley. Energy in this type of valley tends to be stagnant and insular.

Dry valleys are less nourishing than valleys with flowing water. Gullies with their crumbling slopes carry unstable and precarious energy.

Clouds and Mist

Clouds and mist that hug mountainsides are considered to have benevolent energy because they connect the energies of sky and earth. The fleeting moments of rising mist and changing clouds are associated with the rise and fall of breath, or qi, in the land. Such views are consid-

A. Valley formations with negative energy

B. Valley formations with positive energy

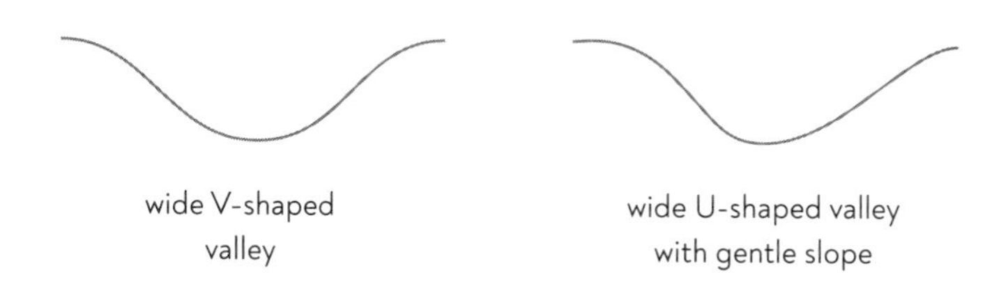

FIGURE. 2. Examples of valley land forms with negative and positive energy.

ered beneficial. However, if you are always fogged in or buried in mist for most of the year, it is no longer beneficial. Instead of being embraced by qi, you will be covered by a lid of stagnant air. Not only will your view be blocked by an opaque dome that is difficult to break out of, but your views will be occluded as well. Some valleys or mountain slopes are covered by clouds or mist most of the year, making such places undesirable to live in.

Natural Waterways

Natural waterways include rivers, lakes, seas, swamps, and marshes.

Fast water carries speedy, restless energy. Examples are white water, choppy seas and lakes (even if it is seasonal), and big waves crashing onto a shore.

Gentle, clear water carries nourishing energy. Examples are meandering rivers, calm seas and lakes, and gentle waves lapping against a shore.

Stagnant water is associated with lethargic and stuck energy. Examples are swamps, marshes, brackish water, and silted water channels.

Some bodies of water are typically stormy. Arctic waters and areas of oceans in the paths of hurricanes are stormy for most of the year. Stormy water carries angry, aggressive, and malicious energy.

If a town is situated along large rivers or is on the coast, the entire community will be affected by the energy of the body of the water.

If the body of water is visible from the building, the building will be affected by its energy whether it is nourishing, stagnant, or destructive.

Human-Made Waterways

Human-made waterways take on the energy of their function and the human activity associated with them. These types of waterways include aqueducts, canals, reservoirs, irrigation ditches, open sewers, and water gardens.

Aqueducts direct water from a source to destinations. In ancient times, aqueducts were elegant structures that were not just functional but brought beauty to a city. One look at the Roman aqueducts will confirm that. Today, water is carried in underground pipelines, but in many small European towns, aqueducts are still functional waterways.

Canals are waterways that function like land routes. They connect sections of towns to each other as well as provide transportation links between cities. Well-maintained canals that are beautified by surrounding gardens, walkways, and elegant buildings carry tremendous uplifting energy. It is no coincidence that residences along the canals of Amsterdam and London are highly desirable areas to live in.

Reservoirs are dammed lakes that carry pent-up energy. The water behind the dam is under pressure and prevented from flowing through its natural course. Moreover, water levels in reservoirs will fluctuate seasonally. When the water level is low, the reservoir walls become cliff-like formations, bringing negative energies of discontinuity to the area. Most harmed are areas near the dam. The dam itself is a steep drop between the reservoir and the area below. Additionally, when water is released through the dam, the rush of water is a torrent that carries destructive energy. Needless to say, it is not advisable to live directly above a dammed lake. Living below a dam . . . well, that's something that you would not even want to consider.

Irrigation ditches are like canals except that they typically run through rural areas and do not have elegant buildings and gardens along their banks. Irrigation ditches are different from rivers in that they do not follow the natural course of the land. Instead, they usually cut through the land to service the nearest fields. When the water level is low or when the ditch is not in use, this conduit will have steep walls. This makes areas with irrigation ditches unfit to build or live in.

Open sewers still exist in some cities. Needless to say, sewers carry waste, and it is therefore undesirable to live in areas in the vicinity of sewers.

Water gardens and parks with water features are human-made structures that carry positive energy. These features are associated with life, movement, and elegance. Buildings that have a view of water gardens are desirable to live and work in.

The effect of waterways depends on two factors: size and distance. If a town is situated within a large network of canals, the entire community is affected by the energy of the waterway. If a waterway is visible from certain parts of the town, only those parts will be affected by the waterway.

Small waterways like ditches, canals, water gardens, and ponds will affect a structure if they are visible from the building. The impact is greatest on buildings immediately adjacent to the waterway. Next are those that are farther away but have a clear view of the water features.

Rock Formations

Rock formations have different kinds of energy depending on their shapes.

In fengshui, rocks that have a defined shape resembling animals, people, or even plants are called spirit rocks. For spirit rocks to affect you, you will need to be able to see them. As with mountains and hills, spirit rocks affect you most if they are directly adjacent to your home. The farther away they are the less impact they will have.

Discontinuities in Land Formations

Discontinuities are dramatic changes between two kinds of land formations. They are associated with drastic and violent energy. Understandably, we should not live close to land discontinuities.

The most dramatic discontinuity in land formation is the cliff. Cliffs are where we find a sharp drop in height between two areas of land. Do not choose a building located at the edge of a cliff, whether you are on the top or at the bottom of it.

Where land meets sea is also a sharp discontinuity between earth and water. If the coastal water is rough, this adds to the severity of the discontinuity. Do not choose a building located right at the edge of water.

How far should you be from a discontinuity before you are safe? Typically, three hundred feet (approximately one hundred meters) should suffice.

The Presence of Animals and Plant Life

Animals are attracted to positive energy and repulsed by negative energy. In Chinese tradition, deer, birds, foxes, wolves, and small ground animals such as rabbits and squirrels are especially sensitive to energy. They gather in areas where there is nourishing energy and avoid places where they can feel harmful energy. If you choose to live in a rural area, take some time to look for the presence of wildlife or ask local residents if there is wildlife in the area.

There are some regions in the world where despite the presence of seemingly hospitable habitat, there is no presence of wildlife. I have visited old battlefields and sites of massacres. Even after a hundred years or more, wildlife still avoids those areas.

Pets are also sensitive to energy. If you have a dog, you may wish to bring her along in your house-hunting trip. I have seen dogs afraid to enter certain buildings and then discovered that someone had died a violent death there.

Plants are also drawn to energy. In an urban environment where there are trees, see if the foliage is healthy or not. In rural areas, see if the trees and grasses are diverse and abundant. Trees with unhealthy foliage are usually associated with places where land energy does not support plant growth. The bottom line is this: if a place does not nourish plants, it won't nourish you.

9

People Energy

THE FENGSHUI OF THE NEIGHBORHOOD

A NEIGHBORHOOD CARRIES ambient, or background, energy. The nature of the energy depends on the types of buildings, land use, the architecture of buildings, street patterns, and the mix of commercial and residential use. Some neighborhoods carry uplifting, fresh, and nourishing energy. Others carry aggressive, fearful, and suspicious energy. I have used these guidelines to help municipalities design neighborhoods. As a renter or buyer, you won't be able to change a neighborhood or design one, but you can pick a neighborhood that is conducive to your health and well-being.

How would you go about figuring out the energy of a neighborhood?

Before you select a place to live, take a walk around the neighborhood and get a feel for how the land is used. Different uses of land bring different kinds of energy into the area. Some uses carry nourishing, peaceful, and harmonious energy. Others carry aggressive, decaying, and unhealthy energy. Second, notice the look and feel of the buildings in the neighborhood. Look at the shape and appearance of architecture. Some buildings radiate peaceful energy; others radiate abrasive and unfriendly energy.

Land Use

The use of land affects the energy of a neighborhood. Ask yourself these questions when choosing a neighborhood to live in:

- Is the neighborhood mainly residential?
- Is the neighborhood mainly commercial?
- Is the neighborhood a balanced mix of residential and commercial?
- Are there parks and green spaces in the neighborhood?
- Where are public and municipal facilities such as schools, hospitals, government buildings, police stations, fire departments, and sanitation stations located?
- Where are utility installations such as cell towers, electrical transformers, and power stations located?

Types of Neighborhoods

Mainly Residential Neighborhoods

Mainly residential neighborhoods are areas that have little commercial usage. There may be a small neighborhood store or a local eating place, but most buildings are residential.

What should you look for in a mainly residential neighborhood?

First, take a look at the conditions of the buildings. Are they well-kept or neglected? A neighborhood with neglected, abandoned buildings will carry energies associated with decay, destruction, and hopelessness. A good neighborhood is one in which people care about the environment they live in, not only for themselves but also for their neighbors. It has nothing to do with whether a neighborhood is affluent or not. Well-maintained nonaffluent neighborhoods can have an uplifted feel and carry benevolent energy.

Second, take a look at the appearances of the buildings. Do they give a "friendly" impression? A neighborhood with buildings that have windows with iron bars or are surrounded by high walls topped with barbed wire or fences with sharp metal points will carry energies associated with unfriendliness, suspicion, and aggression. Again, this has nothing to do with whether a neighborhood is affluent or not. If people surround themselves with unapproachable structures, it is unlikely that they will be approachable, let alone helpful and friendly.

Third, spend some time in the neighborhood during weekends or after work hours and observe the activities of the residents. Do children come out to play? Do adults greet their neighbors and spend time chatting with them? When people in a neighborhood are in touch with each other, an energy of relatedness and even caring is generated. You may wish to speak with residents walking in the area or sitting outside their homes if you are considering living there. Friendly residents would be happy to share their experiences of living there with you.

Lastly, look for the presence of animals. If there are trees in the neighborhood, you should hear the sounds of birds, especially in spring and summer. You may even see squirrels and other small rodents darting around. If you are in a rural neighborhood, the presence of deer, large birds, and small wildlife such as rabbits and rodents will signify that there is life energy in the area. Wildlife is attracted by life energy and repelled by dead energy. If there is no sign of wildlife in an environment that can support wildlife, something is not right.

Purely residential neighborhoods that are uplifted and friendly can be peaceful and desirable for families. However, if you are attracted to atmospheres that are lively, diverse, and vibrant, you may find residential neighborhoods sleepy and boring. In this case you may want to consider living in a mixed residential and commercial neighborhood.

Mainly Commercial Neighborhoods

Mainly commercial neighborhoods are areas where most of the buildings are used for commercial activities with a few residential properties scattered around. Mainly commercial areas are not desirable places to live.

If you reside in an area, there should be residential energy in that area. This means that there should be the presence of people living, not just working there. Residential energy carries domestic energy. Without domestic energy, it becomes challenging to have a nourishing domestic life. This is the reason why most people intuitively avoid living in predominantly commercial neighborhoods, which are typically deserted after work hours. Moreover, the scarcity of residents means that these areas are "socially barren."

If you really want to live in a mainly commercial neighborhood, you need to find out what types of commercial businesses are found in the area. Are they retail spaces, offices, service or hospitality businesses, or industrial companies?

Retail businesses include sellers of food, clothing, electronics, recreational goods, household and business supplies, and so on. Retail brings in customers during open hours, but these spaces will be deserted after the shops close. However, the storefronts of retail businesses typically look dignified because they need to attract customers. Hence there will be no decay or decrepit energy associated with retail storefronts. If you live in a retail business area, while you may not be nourished by social energy after the shops close, you can be assured that there will be no negative energy associated with noxious smells, dirt, and general "untidiness."

Areas with business offices also face a loss of social energy when businesses close for the day. Once the offices close, there will be little human traffic, and the area will feel deserted. However, unlike retail storefronts, which can look colorful even when they are closed, office buildings generally look uniform and uninteresting, without the colorful and unique displays carried by retail shops. If you live in an area dominated by business offices, you will not be stimulated by the environment or feel nourished by it. In fact, you will feel obliged to leave your neighborhood to find social life elsewhere. And if you need to leave your neighborhood because there is no stimulation, then the neighborhood is no longer a neighborhood.

How about a commercial area dominated by services? This will include services like banks, medical and legal services, repair shops, and exercise studios. Neighborhoods dominated by these services will again feel deserted after work hours, possibly with the exception of some exercise studios. The energy associated with commercial services is goal oriented—when the service is completed, there is no more social energy. How many people linger after finishing a meeting with their lawyers or bankers? How many doctors will socialize with their patients when the visit is over? If you live in an area surrounded by services, you will not feel much vibrancy from the neighborhood and will need to leave it to be socially stimulated. Again, this defeats the purpose of a true neighborhood.

What about commercial areas dominated by hospitality services like hotels and restaurants? (Small bed-and-breakfast businesses situated within residences are not included in this category.) Restaurant and hotel districts can be lively well into late hours, but this high degree of vibrancy can bring restless and unstable energy into the area. You can only party so much. Overstimulation will not allow you to have restful energy. You may be attracted to it in the beginning, but you will not be able to sustain this kind of "excitement" for long.

How about an area dominated by industries? These include warehouses, manufacturing businesses, and designated industrial parks. Again, these areas do not carry much "residential" energy. They are deserted after work hours and often generate waste and noise that do not nourish domestic life.

Generally, in commercial zones, employees outnumber residents. This means that the "residential" energy of these areas is weak. "Good" neighborhood energy is generated by the lively and cooperative activity of its residents.

Mixed Residential and Commercial Neighborhoods

Mixed residential and commercial neighborhoods are spaces where residential and commercial uses are interspersed, integrated, and equally represented. An ideally integrated neighborhood is one in which commercial and residential activities mutually support and enhance each other. On one hand, the residential energy brings ambient domestic energy to the residents and livelihood to the businesses. On the other hand, the commercial energy brings vibrancy and diversity for the residents to enjoy. Most importantly, the neighborhood does not feel deserted after the businesses close. People walk their dogs, residents socialize with each other, and the streets are alive rather than deserted. If there is a park or recreation facility nearby, this enhances neighborly energy further. This kind of neighborhood is very attractive and nourishing to live in.

Less ideally integrated commercial and residential neighborhoods are areas where residences and different types of business are haphazardly distributed. The residential and commercial activities don't support each other and seem to coexist by accident. By chance, the residential and commercial energies could enhance each other, but it is not planned and guaranteed. While not ideal, this kind of neighborhood is definitely better to live in than one in which commercial and industrial activities dominate.

The Importance of Urban Design

Urban design is the key to building neighborhoods where a harmonious and supportive interaction can occur between commercial and residential use. In many large cities, we find apartments on top of grocery stores, neighborhood restaurants, and household supply stores. This kind of urban land use is by design and does not emerge haphazardly.

I have worked with high-rise developments in Hong Kong and Singapore where entire communities are built so that residential and commercial use complement each other. Typically, the upper levels of the high-rise are residential, and the lower three or four levels are commercial, consisting of grocery stores, clothing and household-supply shops, restaurants, services such as hair salons, laundry facilities, child-care centers, exercise studios, and even a small movie theater. This kind of arrangement allows residents to socialize easily, not to mention the practicality of taking care of your domestic needs indoors during severe weather.

From the fengshui point of view, this kind of mutually supportive residential and commercial mix generates a friendly and cooperative energy among people. Neighbors run into each other frequently, and spontaneous human cooperative activity can arise.

Community Space

Parks and Recreation Spaces

Parks are open spaces, whether they are soft- or hard-scaped. Having open spaces among buildings can facilitate the flow of energy, calm down fast energy, and slowly release pent-up energy. In a simple way, we can say that parks allow a neighborhood to "breathe." This "breath" is qi, or energy.

Softscape parks are typically green spaces filled with trees, bushes, grass, and ponds. Most green-space parks are restricted to pedestrian and bicycle traffic. Wildlife is typically more abundant in these areas. Generally, people come to these parks to enjoy nature, and their activity tends to be more relaxed and leisurely. Green space therefore radiates a soft, relaxing energy to the neighborhood around it.

Hardscape parks are dominated by concrete grounds such as outdoor ball courts, skate parks, swimming pools, and indoor recreation facilities. This kind of park is typically dominated by activity. "Activity" parks will radiate vibrant and intense energy to the neighborhood. However, once the activity is over, people will leave and the area will become deserted.

The Town Square

The town square is an interesting outdoor space that can gather positive energy, dissipate negative energy, and bring human energy into a space.

Originally designed as a market place in both Asian and European cities, town squares (or plazas) are unique in integrating residential and commercial energies where they can mutually enhance each other. This kind of open space encourages pedestrian traffic and helps slow down speedy energy often associated with cars. The modern imitation of the traditional plaza is the shopping center. However, one big difference between the two is the presence of parking lots. You can approach a plaza by walking there leisurely. In contrast, you have to negotiate the stress of finding a parking space before you can enter a shopping center.

Markets

Outdoor and indoor markets can bring lively domestic energy into a neighborhood. Farmers markets, pier-side markets, and food stalls are areas where people can mingle, create vibrant and diverse energy, and promote a healthy mix of residential and commercial activities.

Public Facilities

Public facilities include schools and colleges, hospitals, libraries, government buildings, police stations, fire departments, sanitation stations, transportation terminals, parking garages, cemeteries, correction facilities, and public utilities installations. Some public facilities carry positive energy, some carry negative energy, some are neutral, and some will depend on their architecture.

Generally, schools, colleges, and libraries carry energy associated with learning and scholarship. Schools for young children and daycare centers carry energy associated with the vibrancy of youth. These types of facilities bring nourishing, intellectually stimulating, and youthful energy into a neighborhood, making it an area desirable to live in.

Hospitals often carry the energy of illness and death. Cemeteries are definitely associated with the energy of death. Therefore it is not desirable to live near hospitals or cemeteries. The exceptions are clinics and community health centers that focus on preventative medicine and promote health and mental well-being.

Sanitation stations carry the energy of decay. These include dumps, junkyards, garbage-processing stations, and even recycling plants. The issue is not whether such stations are hygienic or operate according to health standards. The issue is that they are surrounded by garbage, which is composed of discarded or decaying items. No one wants to live near a garbage dump.

Transportation terminals and parking garages carry the energy of movement that is restless and speedy. The energy in these structures is constantly moving and never resting. People and vehicles are always going somewhere. Needless to say, living near these facilities will not give you restful energy. Even hotels linked to these facilities or that are in their vicinity will not give you a good restful sleep. For this reason, airport hotels do not make relaxing resorts.

Fire departments and police departments are facilities that should ideally carry an energy of protection. However, many police stations exude an atmosphere of fear, violence, and authoritarian power. The architecture and human activity around a police station can contribute to the energy the facility radiates. Police stations with large heavy doors and small windows and no way to view its interior will generally carry an energy of fear, suspicion, overwhelming authority, and lack of outreach toward the community. This architecture affects the behavior of those who work inside and those who live nearby.

Fire departments are associated with "emergency" energy. While their goal is to rescue and help, the fact that they are always responding to emergencies makes the energy around a fire station fast and stressful. In addition to the annoyance of sirens and speeding emergency vehicles, living close to a fire station can make someone feel apprehensive and panic prone.

Prisons and correction centers carry violent energy. It is definitely not desirable to live in the vicinity of these facilities.

Other facilities that carry disruptive and destructive energy are power plants, communication towers, and large collections of satellite dishes associated with broadcast stations. These structures not only carry magnetic energy that is harmful to health but also create vortices of energy that destabilize livelihood, health, and relationships. Energy generated by these large structures will overwhelm any positive human energy that circulates in the area. Moreover, since most of these facilities have pointed structures, people in the buildings nearby will feel attacked by them.

Roads

Roads include city streets, intercity and rural roads, high-speed thoroughfares with entrance and exit ramps, paved urban roads, and unpaved rural roads. In fengshui, roads are conduits of energy.

Types of Roads

Fast roads carry fast energy. The speed of the traffic is like fast-flowing water. If you live on a road with fast traffic, benevolent energy will not have a chance to linger and settle. Moreover, the fast traffic will make the energy in the area speedy, restless, and unstable. It is difficult to build good neighborhood energy along a fast road because fast traffic does not encourage leisurely pedestrian activity that is conducive to neighborly interactions.

Slow roads carry slow energy. If you live on a road with slow traffic, benevolent energy will have a chance to settle and enter your home. In addition, the slower traffic will increase the chances of building a cooperative neighborhood. People will be more likely to linger on the street and socialize. Parents will feel that it is safe to let children play outside. Generally, slower roads bring leisurely energy that encourages pedestrian traffic. No one wants to take an evening stroll or walk their dog along a road with fast traffic. Many towns have tried to slow down traffic in residential neighborhoods by putting in speed limits and speed bumps. This is a good step toward creating a safe and cooperative neighborhood.

While slow roads carry less frenetic energy, you do not want to live on a deserted road. Some kind of traffic is needed to bring energy into an area. You do not want to live in an area with abandoned buildings where there is almost no traffic.

Wide roads generally mean more traffic. A four-lane road will carry more traffic than a two-lane road. When there is more traffic, there will be more restless energy. More vehicles also mean more energy is swept through with little chance of settling. It is definitely not desirable to live on a wide road. The worst case is a wide road that has busy and fast traffic.

Narrow roads do not necessarily mean less traffic. In fact, narrow roads such as one-lane roads and narrow streets in old city centers, especially in Europe, are associated with choked energy regardless of the amount of traffic. Narrow streets are like canyons, especially when they are flanked by tall buildings. Sunlight cannot enter these streets, and it is always dark and foreboding.

Congested roads are choked by nonmoving traffic. Energy is locked up because nothing is moving smoothly. Living along congested roads will lead to your life being congested and choked as well. Health, livelihood, and relationships will be blocked. In other words, there is a high likelihood that things will go nowhere in your life. This is definitely not a good way to live, so don't live on roads that are habitually congested.

On steep roads energy rushes down uncontrollably. Positive energy from the top will zip by and never enter your space, and positive energy from the bottom won't be able to climb up to enter your space either. Moreover, steep roads carry fast energy, making a neighborhood speedy and restless. The situation is worsened when the road is winding and/or when there is heavy traffic. It is definitely not desirable to live on steep roads.

The best kind of road to live on is somewhere in the middle—roads that are not too narrow, not too wide, have moderate traffic, and are not deserted.

Area Road Patterns

Road patterns refer to how roads are configured. These configurations can generate and carry different kinds of energy—benevolent, nourishing, aggressive, confusing, dull, creative, and even deadly.

Straight roads carry energy with little chance of slowing and entering buildings. Energy carried by straight roads, even if the traffic is slow, is usually dull. Straight roads typically have one purpose: they are thoroughfares routing traffic from one point to another. If you want to get from one location to another quickly, you will travel along the straightest path.

On the positive side, negative energy from garbage, dead and decaying matter, and even hazardous materials traveling on these roads will not linger long enough to enter buildings. On the negative side, positive energy associated with wealth, health, and well-being will pass by without a chance of entering the buildings. The only time positive energy can spread into the neighborhood is during a parade. The slow-moving parade traffic allows the uplifting energy of the celebration to spread into the buildings lining the street.

Winding roads typically slow down traffic, especially in gentle terrain. A meandering road winding through a neighborhood with parks and waterways not only slows traffic but also contributes to the beauty of the surroundings. Winding roads make neighborhoods interesting—you never know what will appear around the next turn. I always look forward to driving leisurely along gentle winding roads and getting surprised and delighted by what appears around the corner.

There is one type of winding road that is guaranteed to carry destructive and even deadly energy. These are the narrow roads that wind down steep slopes with a sheer drop-off on one or both sides. Buildings perched on the steep sides of the roads are not only buffeted by fast aggressive energy, but health and livelihood can be destabilized by the energy of traffic moving through fast winding roads. Steep roads in themselves already carry fast energy, and traffic moving along these roads will amplify the severity of the rush of energy down the road—regardless of their speed of movement.

Grid-like road patterns carry no-nonsense energy that is uninteresting. The neighborhoods are laid out in squares, each block about the same size. Energy in neighborhoods with grid-like roads is typically not creative. If you want to be artistic or even bold in experimenting with new ideas, this type of neighborhood is not for you. However, if you want a stable, slow-paced, and unadventurous life with little to no surprises, this is where you would want to live.

The best layout of road patterns is a good mix of grids and meandering roads. For example, a neighborhood made of square blocks of residential units punctuated by roads meandering around green space is the best of both worlds. There is regular energy that stabilizes domestic life, especially for families, as well as occasional creative energy carried by gentle winding roads through pleasant scenery.

Maze-like road patterns are the worst kind to have in a neighborhood. The flow of energy in this road pattern is blocked and confused. You go around in circles and get nowhere. You can determine whether an area of town has a maze-like road pattern by driving from one corner to another. If your route is convoluted, then you have entered a maze-like area. If you feel lost and disoriented, this is another sign that you are in a maze. A good way to determine whether an area has a maze-like road pattern is to look at a street map.

Road patterns can literally define the boundaries of a neighborhood. When a street pattern changes, we enter a different neighborhood.

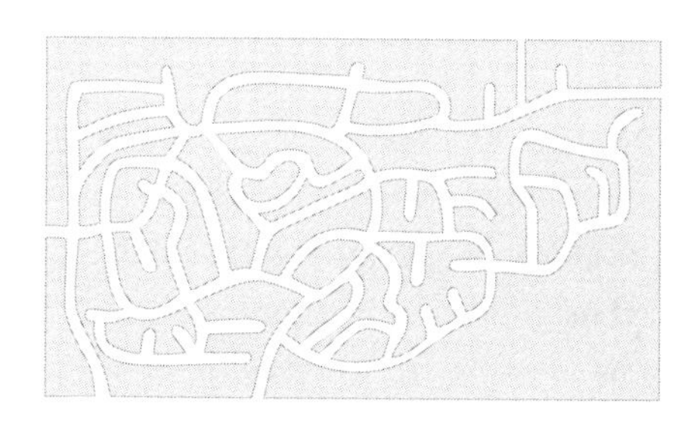

FIGURE 3. Maze-like street pattern.

FIGURE 4. Street patterns defining neighborhoods.

Local Road Patterns

Local road patterns affect the energy of a building situated there. You can have an area road pattern that is clear and simple, but certain specific locales along the road may be harmful. The following are locations that you don't want to live or work in. If you live in these locations as a resident, you will take the full brunt of the negative energy. If you work in a building in these locations, the effect of harmful energy will be shared among those who work there.

A road that bends sharply and almost loops back onto itself is called a "noose" road. Buildings located within the noose will have their energy strangled and confined. Living in this location will make you feel imprisoned and choked. A **T-junction** is exactly what it describes. The junction is T-shaped, with a straight road running into another road at a right angle to it. A building located at a T-junction will be slammed by energy from incoming traffic driving straight toward it. The energy at a T-junction is associated with accidents, collisions, and constant onslaught of destructive incidents. You feel like you are being hit by a battering ram all the time.

A **V-junction** is formed by a road making a sharp turn like a switchback. A building located where the V is pointing will receive sharp, destructive energy. You'll feel like you are being constantly punctured by a spear. Life plans, relationships, livelihood, and health will be attacked and

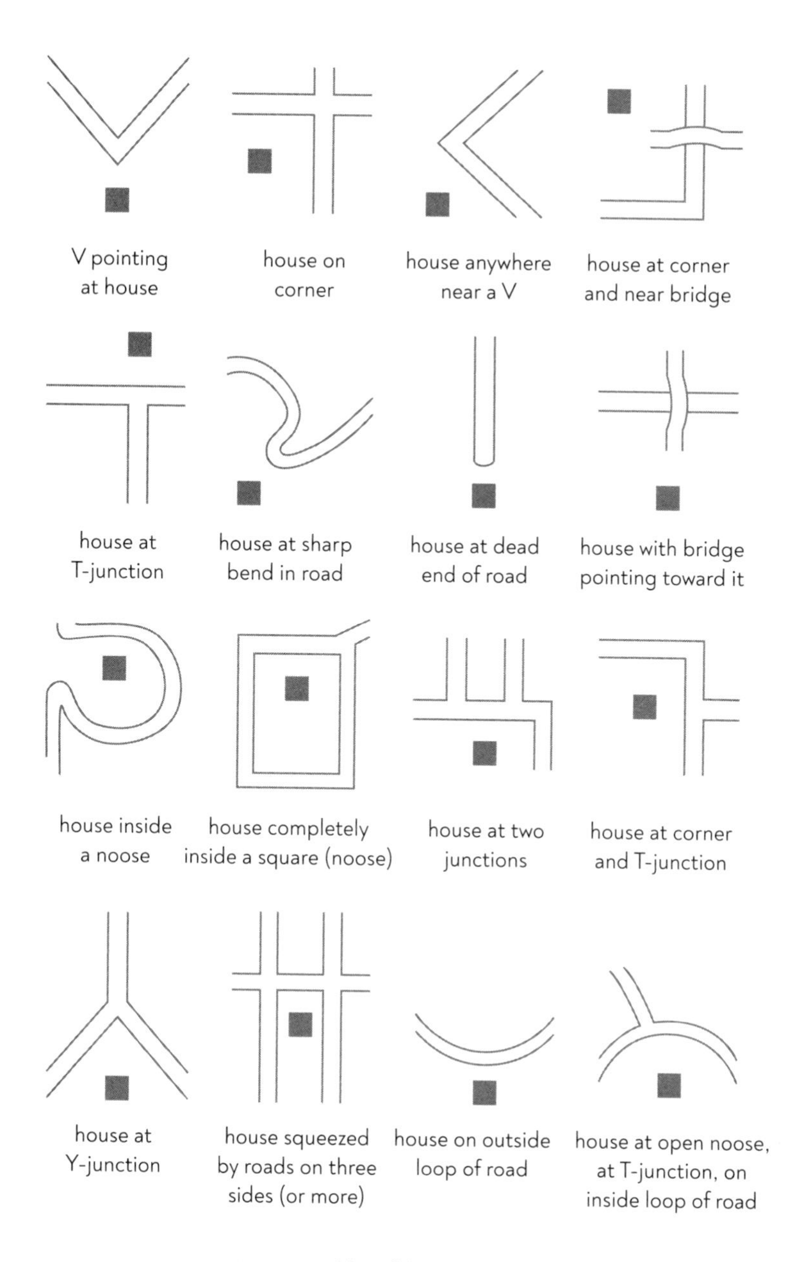

FIGURE 5. Harmful road patterns.

will have little chance of reaching fruition. The effect is the same for buildings located at a sharp bend in the road.

A **Y-junction** is formed by a straight road branching into two directions. A building located at a Y-junction has the same problems as one located at a T-junction. However, the Y-junction also has the additional negativity of a noose. This is especially true when the intersection at the Y-junction is constricted.

A **U-junction** is a Y-junction with roads branching off circularly. The effect is the same as that of the Y-junction.

A **dead end** is where the road ends. You are literally and energetically at the end of the road and can go no farther. Energy at a dead end is associated with a lack of choices, options, chances, and future—not exactly a good situation to be in.

A **convex bend** is where the road bulges out. Buildings located at the bulge will be buffeted by energy of traffic approaching the bulge. The negative effect is not as severe as a T- or Y-junction, but the annoyance of constantly being buffeted by obstacles makes it an undesirable place in live.

Corners are intersections where vehicles make sharp turns. Sharp turns are associated with a sudden and sharp change of energy. If you live or work in a building located at a corner you will feel your work and life energy making sharp and sudden turns without warning. The busier the intersection, the more sudden and speedy the changes will be. If there is a traffic light at the intersection, there will be the added negative effects of irritability, impatience, and even aggression as vehicles try to run red lights or speed through stale green lights. It is no coincidence that accidents occur most frequently at intersections with traffic lights.

In **traffic circles** or rotaries, vehicles change direction of travel by entering and exiting a circular route. Although vehicles usually slow down in a traffic circle, this type of road pattern creates an energy akin to a whirlwind. Living in the vicinity of such a vortex is like living on the edge of a tornado. You will not have much time for peace and quiet in your livelihood, relationship, or quality of life.

The **cul-de-sac** is a common road pattern in many residential areas. In a cul-de-sac, a road enters a circle where buildings are situated along the circumference.

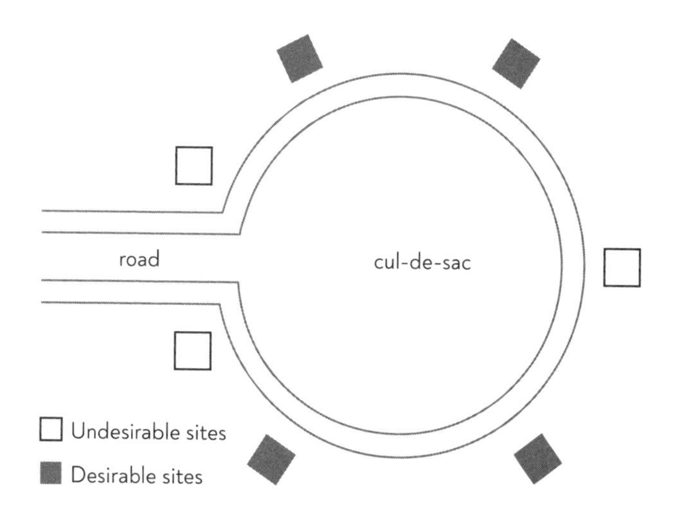

FIGURE 6. The cul-de-sac.

The area where the road enters the circle is a zone of energy discontinuity. The discontinuity is defined by two vastly different kinds of energy coming together—the straight road and the circular driveway. Buildings located in these areas usually encounter energetic turmoil associated with this discontinuity.

A building positioned directly opposite the entrance to the cul-de-sac will receive the brunt of harmful energy from the road pointing straight toward it. This negative effect is similar to a building located at a T-junction.

The best sites along the cul-de-sac are those located away from the entrance and not opposite the access road. Buildings situated on these sites will receive the benefits of positive energy associated with circular structures and be nourished by energy tamed and collected in the cul-de-sac.

Train Tracks, Airport Runways, Bridges, Tunnels, and Elevated Highways

Railways and tram tracks, airport runways, and highways literally cut into neighborhoods and divide them. These traffic thoroughfares carry more destructive energy than roads because the mass of negative energy carried by trains and airplanes far exceeds that of ordinary road vehicles. Railway tracks and runways often cut through land indiscriminately, forming barriers between neighborhoods. These barriers often make residents feel trapped and even fearful of crossing the barriers into other neighborhoods. These kinds of communities often suffer poverty, inequality, and even discrimination in social services and protection. Those living in buildings next to these traffic structures receive the brunt of aggressive, fast energy detrimental to livelihood and well-being.

Bridges and elevated roads are built with large concrete slabs and steel girders. These structures carry energy associated with overwhelming strength that will crush anything beneath them or in their vicinity. Residents and workers in buildings located near bridges and elevated roads will feel crushed by heavy burdens and buffeted by fast, unstoppable energy.

Highway exit and entrance ramps not only carry fast and frenetic energy, they are also junctures where vehicles change speeds when entering or exiting a high-speed road. This creates an energy of discontinuity and unpredictability. Drivers are typically stressed, and this energy of stress and anxiety will spread into buildings located near the ramps. It is no coincidence that accidents are more likely to occur when vehicles merge onto high-speed roads or exit from them.

Tunnels are one of the most destructive traffic structures. First, they cut through mountains or beneath the earth and are an example of disregard for the land and indigenous people and wildlife who inhabit those areas. Expedience, convenience, and greed often motivate the building of tunnels. Many tunnels are built solely for the purpose of transporting goods for profit, mining the land for minerals, and moving people quickly to destinations for self-centered interests. Second, where the tunnel enters or exits the mountain, there is negative energy associated with discontinuity formed by the transition from "sky" energy to "underground" energy. Third, traffic inside the tunnel will destabilize energy above. For all these reasons, buildings located near tunnel entrances or that are on top of them are undesirable to live and work in.

10

Architectural Energy

THE FENGSHUI OF BUILDINGS

BUILDINGS CARRY and radiate energy. The energy is defined by the building's architecture, which includes its appearance, its shape, and the materials it is made of. These architectural details make up the "look and feel" of a building. Some buildings carry and radiate a sense of calmness, some radiate aggression and intimidation, others radiate fear, and some even radiate a false sense of security.

Appearance of a Building

The appearance of a building determines the effects fengshui has on the building itself and the structures nearby. When looking for a place to live, pay attention to the look and feel of the building itself as well as those around it.

Aggressive-Looking Buildings

If a building "looks" aggressive and intimidating, then the energy it carries and radiates will be aggressive and intimidating. People living and working in this type of building will take on the aggressive energy carried by the structure, and those who live in full view of this type of building will be on the receiving end of this energy.

Features of Buildings with Aggressive Fengshui

- An ax or spear-like appearance
- Shiny reflective surfaces
- Destructive objects like sculptures of weapons integrated into their architecture
- Spiky domed roof structures
- Knife-like edges
- Arrow-like dormer windows
- Facades that consist of massive stone pillars
- A fortress-like appearance with small windows and dark facades
- Massive stone facades
- A battering-ram-like appearance

Buildings with Unstable Energy

Some buildings are unstable energetically. While these buildings may be structurally sound and safe, their energy is unstable. The instability is manifested in the appearances of these structures. If a building "looks" like an unstable structure, then its energy is unstable.

Features of Buildings with Unstable Energy

- Thin supporting pillars with space underneath, such as buildings on top of garages, structures built on pillars sunk into the hillside, or structures built on elevated platforms
- Upper levels that are larger in size than the lower levels, making the building appear "top heavy"
- Irregular levels that are randomly different sizes
- Sections that are cantilevered out into space
- "Holes" in the center where some levels of the high-rise are "missing"
- Thin stilts as supports

Special case: Some buildings need to be built on platforms supported by stilts for safety. An example are buildings in flood-prone areas. This seeming "instability" can be mitigated by putting a trellised wall along the sides of the ground level, making the building appear as if it is sitting on a "walled" structure. I have used this architectural feature in many buildings in flood-prone areas in Louisiana, coastal Florida, Indonesia, and Thailand to mitigate the energy of instability.

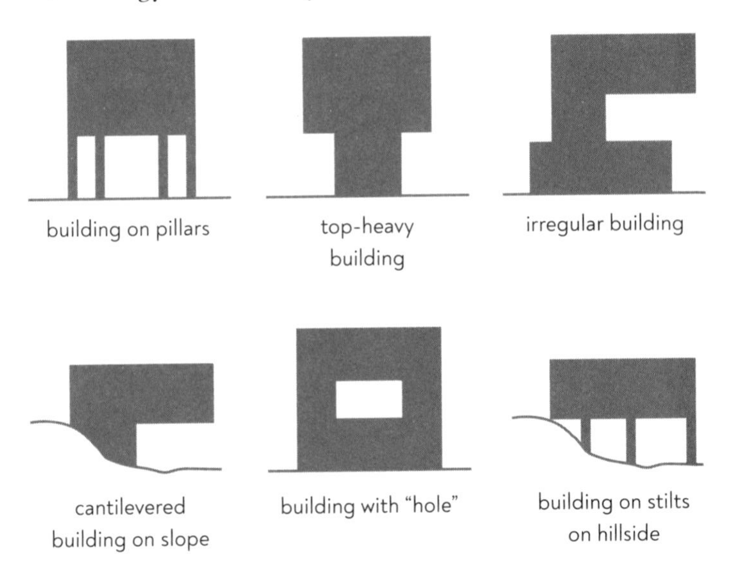

FIGURE 7. Examples of buildings with unstable energy.

Surface Structure of a Building

The surface structure of a building also affects the ease of energy flow in and around it. If the facade or walls of the building are smooth and uniform, the energy inside will flow evenly. If a building does not have a smooth surface, the energy inside will suddenly burst into one area or get squeezed into another, making it difficult for residents to deal with violent swings of energy as they go from one section of the building to the other.

Features of Buildings That Lack Smoothness

- Sections bulging out or pushed in
- Facades composed of large irregular stones
- Windows positioned irregularly
- Balconies positioned irregularly
- Facades composed of a patchwork of brickwork or stonework.

Friendly Looking Buildings

While some buildings carry negative energy, there are others that carry positive or benevolent energy. This positive energy is manifested in their appearance. If a building "looks" friendly and approachable, then its energy is also friendly and benevolent.

Features of Buildings That Carry and Radiate Benevolent Energy

- Elegant features such as gardens integrated into their architecture
- Art and decor that are pleasing to look at
- Smooth facades
- Symmetrical features
- Abundant natural light

Architecture That Carries and Radiates Neutral Energy

Neutral energy is energy that neither benefits nor harms. If the appearance of a building appears uninteresting and forgettable, then it is likely that it has neutral energy.

Summary of the Appearance of a Building

In summary, if you look at a building and feel intimidated by it, you should not live in it or near it. If a building looks pleasant and elegant, this is a desirable place to live in or live nearby. If nothing feels interesting about a building, the building is said to have neutral energy. Its look and feel may be boring, but it would not affect you positively or negatively should you live in it or near it.

Shape of a Building

Energy is also affected by the shape of the building.

Regularly Shaped Buildings

Buildings with a regular shape (square or rectangle) carry symmetrical energy. Energy inside this kind of building will be able to move equally to and settle in all areas. There will be no one area where energy, positive or negative, will dominate.

Irregularly Shaped Buildings

In irregularly shaped buildings, the flow of energy will be unequal and even chaotic. Some areas inside will have more energy than others. In an irregularly shaped building, positive energy tends to be more benevolent and negative energy tends to be more destructive.

At first glance, round-shaped buildings may feel benevolent, but their roundness makes it difficult to partition the space into rooms. When round-shaped buildings (like yurts) are partitioned, the rooms tend to be irregular, making the energy captured inside these rooms also irregular. If you want to live in round-shaped structure, you should have an open floor plan with as little partitioning as possible.

In triangular-shaped buildings, energy is trapped in the apices of the triangle. Trapped energy is stale energy. Moreover, triangles are shaped like arrowheads, and therefore can carry sharp and destructive energy. Buildings with a triangular structure pointing at them will be on the receiving end of destructive energy. Those who live and work in triangular-shaped buildings will be influenced by the sharp energy of the corners, which will make them feel constricted and irritable. It is also difficult to place furniture in a triangular space. For all these reasons, it is best to avoid living in or near a triangular-shaped building.

Energy in octagonal buildings is defined by the eight points of the octagon. Corners formed by octagons are not as sharp as triangles, but they create space that is sharp and pointed. Similarly, the energy in an octagon is not as sharp and destructive as a triangle's, but it will create a space that is uncomfortable. Rooms created by partitioning an octagon tend to be irregularly shaped, giving the energy a feeling of irregularity. Octagons have problems similar to round spaces, making room partition difficult. Moreover, the sharp corners can introduce abrasive energy into the space. For these reasons, it is best to avoid living in an octagonal space.

The shape of a building molds the energy that flows inside it. Buildings with irregular shapes will carry irregular energy. By irregular energy we mean that energy flowing inside that building will change dramatically from one area to another. Thus, when you walk from one room to another, you will feel a change in the mood of the space, for example, from peaceful to aggressive or from excitable to dormant. Buildings with an irregular shape often lead to residents having wild mood swings.

Features of Buildings with Irregular Shapes

- A zig-zag layout
- A collection of shapes such as squares, triangles, circles, etc.
- "Chopped-up" sections that are linked by breezeways and corridors

Roof Features

The shape of the roof, the structure of roof lines, and the architectural features on a roof can affect the energy inside a building. Roofs cover the entire building, and their shape and contours act like a lid enclosing the energy within. Buildings with complex roof lines carry complex energy. Life in the modern world is already complicated as it is. There is no need to add more complexities to our lives with complex energy coming from uneven and complicated roof lines.

Architectural features on a roof also introduce energies associated with those features. Dormers that form bulges on the roof disrupt the smooth flow of energy. They act like obstacles, preventing energy from flowing evenly across the roof and downward into the building. Triangular dormers bring abrasive energy into the house; square dormers divide space and introduce energies of "turf war" inside the house. Most problematic about dormers is that they form blade-like walls inside the building. Placing furniture, especially beds, under the "blades" will have a negative effect on the health of the occupant using that space.

The only kind of dormers that are viable are shed dormers. Shed dormers look more like a story that was added to the building. The "blades" formed by the walls of the dormers are typically close to the side of the building, making it easy to hide them by building closet space around them.

Older houses and houses with attics transformed into living space tend to have square dormers. Mid-century houses tend to have triangular dormers. Modern houses tend to have "mock" dormers, meaning that the dormers are just surface architectural features "stuck" onto the roof. Not only do they have no function, but they also often leak, which can be costly to repair.

How can you deal with dormers if you already live in a building that has them? First, it is not difficult to replace the triangular and square dormers with a skylight window. A more costly but better solution is to turn the problematic dormers into a large shed dormer. You increase the living space and ceiling height and improve ventilation.

Building Materials

Building materials themselves carry energy. This energy is based on the ancient Chinese theory of the five elements that is widely used today in fengshui, Chinese medicine, and Taoist cosmology. The five elements are metal, wood, water, fire, and earth. Metal energy is sharp and incisive; wood energy is strong, heavy, and plodding; water energy is flexible but fragile; fire energy is volatile and aggressive; earth energy is stable but dormant.

In fengshui, each element has different pros and cons in the way it affects the kind of energy that is contained in a space built with those materials. The most common materials found in buildings are: wood, metal, rock and stone, glass, and mixed materials. In fengshui, a building is said to embrace an element if the structure is built primarily from that element and if that element dominates the appearance of the building. For example, a log cabin has dominant wood energy because it is not only built almost entirely from wood but also carries a strong appearance of wood in its beams and pillars. Even the ceiling and walls are covered with logs.

Structures that are primarily built of wood will look and feel strong externally, but the energy will feel heavy and possibly claustrophobic inside, especially if there are large, thick wooden beams in the ceilings and along the walls. Living inside a wood-dominated building will pull your energy down. You will also feel "burdened" by logs pressing against you. If you have no choice but to live in a wood-dominated building, make sure your bed, dining table, and favorite couch are not under a large beam.

Structures that are primarily built of metal will look sharp and impenetrable externally. The metal "shell" of the building will enclose an energy that tends to have a sharp and cutting nature. Metal-clad buildings also reflect energy. While it will reflect incoming negative energy back outside, it will make it difficult for benevolent energy to enter as well. If you live in a metal-dominated building and don't want to be overly sharp and cutting, the best solution is to cover the metal with wooden siding or paint it with nonreflective paint to lessen the metal-like appearance.

Structures that are primarily built of rock and stone will look strong and formidable. It does not matter if the stones are natural or faux. It is the appearance that carries the energy. Stone-dominated buildings radiate the feeling of a fortress or castle. You may feel protected and impregnable, but the energy radiating from a stone-dominated structure is not friendly or welcoming. The building is sending a message, "Keep out" or "Enter at your own risk." If you wish to soften the stone facade, you can place awnings and plants around the building, making the exterior of the building feel "softer" so it will appear more friendly and welcoming.

While rock and stone carry strong earth energy, bricks and adobe carry soft earth energy.

Structures that are primarily built of glass facades are getting more common these days. Buildings with floor-to-ceiling windows and doors of glass are considered glass dominated. In fengshui, glass is considered water by nature because its clear surface resembles water. Glass may allow energy to flow through unobstructed. However, this means that both positive and negative energies can enter a building indiscriminately. Moreover, energy accumulated inside a building can also flow out. In China there is a saying about glass: Health and money can come in but will go right back out. Glass is not a strong material energetically. Unlike wood or stone or brick, it offers little protection. Ideally, bedrooms and offices should not have large floor-to-ceiling windows. You can protect a space and regulate the flow of energy in and out of a room with large glass windows using window coverings and low furniture placed along the glass windows.

While fire is not a building material, the element of fire is carried in certain types of materials. These include fiery-red brick, red walls and siding, and red tiles. Structures built primarily of these types of materials will carry and radiate fiery energy. Those living and working in these buildings will tend to have fiery and volatile tempers. The easiest way to mitigate a building with too much fire element is to paint the walls or siding a blue color. Red bricks can be painted white or a beige color. Once the red color disappears, the fiery element will be gone.

What is the best kind of material to use in building a house? Ideally there should be a balanced mixture of diverse materials. Fengshui is about balance. The way to maintain balance and harmony in a structure is by not having too much of one element. If there is a balance of elements in the building materials, the energy will be balanced and harmonious.

Types of Residential Buildings

If you had a choice of the type of building you would like to live in, what would it be? A single house, a duplex or a row house, an apartment? Convenience may be a factor in your choice, but these structures have different fengshui. It may be wise to know the kind of fengshui you will be living with before you make your final choice.

Single Homes

A single home by definition is not attached to any other structure. First, because it is freestanding, there are no shared walls with a neighbor; therefore, energy from neighbors cannot spread or leak into the space. People have different energies, and while your neighbors can be friendly and even helpful, their energies will not be identical to yours. The diversity of energies make us unique individuals, but it would not be pleasant to have the life energies of others intrude into yours. Secondly, a single home has space around it, allowing the energy inside to radiate and spread. It is as if you have room to stretch and breathe without obstruction. Thirdly, the space around a single home allows you to enhance your home with gardens and outdoor living areas. The enhancement can increase positive energy entering and surrounding the home. Of course, an inconvenience is the added work needed to maintain a single home. Hopefully, you will weigh the inconvenience against the advantage of the fengshui of a single home.

Duplex and Row Houses

A duplex consists of two houses sharing one wall. A row house is a structure that shares both walls with a neighbor. In a duplex, energy from one unit will spread to the other. Both positive and negative energies will leak indiscriminately. We are not just referring to hearing quarrels through the walls but actually feeling the negativity of the conflict. Because only one wall is shared, a duplex home will allow some degree of freedom in spreading your energy outside.

A row house is more constricted energetically than the duplex. This is because both walls are shared with neighbors, with the exception of the corner unit. First, energy will spread from both neighbors into your space, and it will feel like you are being invaded by unwanted guests. Secondly, the space inside a row house is dark because there are no windows on either side of the home. The only light entering the space is from the front and the back. A dark space typically generates lethargic and sleepy energy. No one likes living in darkness. Thirdly, a row house will not allow you much chance to spread your energy. If the row house is long and narrow, this adds to the feeling of being constricted and trapped.

The main convenience of living in a duplex or row house is that they require less work and cost to maintain. Before you choose a duplex or row house, weigh the convenience and cost against the disadvantages in the feng-shui of these homes.

Apartments

Apartment living has become part of the culture of our times. Some people do not like the slow energy of life in suburban and semirural areas and want the excitement and diversity of living in a city center. If you want to live in the center of a city, you will be limited to choosing an apartment (or condominium).

Energy in all-residential apartments is different from energy in apartments with mixed commercial and residential units. In all-residential apartment buildings, the lower level is typically reserved for the garage or recreational facilities. In mixed commercial and residential units, the lower levels can consist of parking space as well as commercial units.

In apartments with only residential units, the energy in the building is influenced by the energy of all the residents. The primary impact is from neighbors that are adjacent to you and share a wall with your unit. Secondary influences come from the units directly above and below you and from across the hall. If you want to live in an apartment, the best option is to be in a building with fewer stories and units. The most ideal apartments are those that have one unit per level and are less than ten stories high. Penthouse units are only affected by energy from those units below them. However, because they are on top of the building, protection is compromised by stormy energy that can descend from above. Moreover, since penthouses are located on top of tall buildings, these units tend to bear the brunt of attacks physically (from weather) and energetically from aggressive architecture nearby.

In apartments with mixed commercial and residential units, the energy in the building is influenced by the energy of both the residents and the businesses. The effect of energy from the residents is the same as that of apartment buildings that are purely residential. However, there are now the added effects of energy coming from the businesses. The impact that businesses have on residences depends on two factors. The first is the distance between the apartment and the business. The second is the type of business.

Residential units directly above a business will feel the most impact from the activity of the business. The farther you are from the commercial levels, the lesser the impact.

ENERGY OF NEARBY BUSINESSES

The type of business and the activity associated with it can carry positive energy or negative energy. Again, the impact of the energy from a business is greatest on apartment units directly above it. Apartments higher up will receive less impact.

Many types of businesses carry positive and benevolent energy that can enhance and enrich apartment living. For instance, daycare centers are not only a convenience to working parents, but children's activity also brings fresh and active energy into a space. Children play, socialize, and communicate with each other with innocence and happiness. It is uplifting to experience the energy of young people. Salons, spas, and health clubs are not just businesses that serve the well-being of their clients but are often spaces where social communication and caring are facilitated and friendships are made. These businesses also build community well-being and bring the energy of cooperation, friendship, and social well-being. And businesses such as art studios and libraries can enrich a community by carrying creative energy that brings culture and its appreciation into an area. Living directly above any of these types of businesses can make residents feel socially and culturally enriched.

There are also many types of businesses that carry negative energy that is disruptive, restless, or stressful.

For example, busy restaurants, grocery stores, laundry services, and general retail businesses. The disruptive energy generated by these businesses simply comes from the fact they are always busy and often carry frantic energy. Shoppers are trying to find goods and are frustrated when they are unable to. Employees are trying to stock goods while dealing with irritable and impatient customers. Most of the disruptive energy comes from the sheer amount of people entering and leaving these businesses. Additionally, shoppers and employees carry their own energies into the space. As a result, we have a cacophony of energies swirling around. Before one kind of energy can settle down, another kind of energy enters. Residents living immediately above these businesses will feel chaotic, unpredictable, and topsy-turvy energy entering their space. Even when the businesses close for the day, residual energy from the daily activity can continue to seep upstairs.

Offices of professional services carry an energy that is specific to their service. For example, medical offices can carry the energy of ill health. In our culture and times, while preventative medicine is now more prevalent, most people only visit the doctor's office when they are sick. The energy of lawyers' offices is associated with unresolved issues, conflicts, and the fear of future legal problems. The goal of financial services may be to help clients improve their finances, but financial energy is volatile by nature. Fortunes come and go; good times give way to bad times and vice versa. Living directly above these kinds of professional offices is definitely not desirable.

Some businesses are associated with aggressive energy. These include gaming arcades and combat-oriented martial arts studios.

If you want the convenience and diversity of city living experiences associated with apartments, the best choice is a unit that is not too high and is several levels removed from commercial usage.

LOCATION OF THE APARTMENT UNIT IN THE BUILDING

Some units in apartment buildings have better fengshui than others. Let's look at the fengshui of different unit locations within an apartment building.

Most people want corner units because there are more views. If the views are pleasant, positive energy can enter the unit from multiple sides. However, if there is negative energy in the surrounding area, that can also enter the unit from multiple sides (see chapter 11, "Object Energy," on page 88 for more information on how surroundings can affect a building's energy).

Apartments that are sandwiched between units on both sides will be vulnerable to energy leaking in from two neighboring units. Because only one side of the unit has a view, it is less effective in welcoming positive energy. However, if the incoming energy is negative, this type of unit will be better protected than the corner unit.

The location of elevators and stairways is an important consideration when choosing an apartment. A unit directly adjacent to the elevator will be buffeted by the rush of upward and downward movement. Since elevators

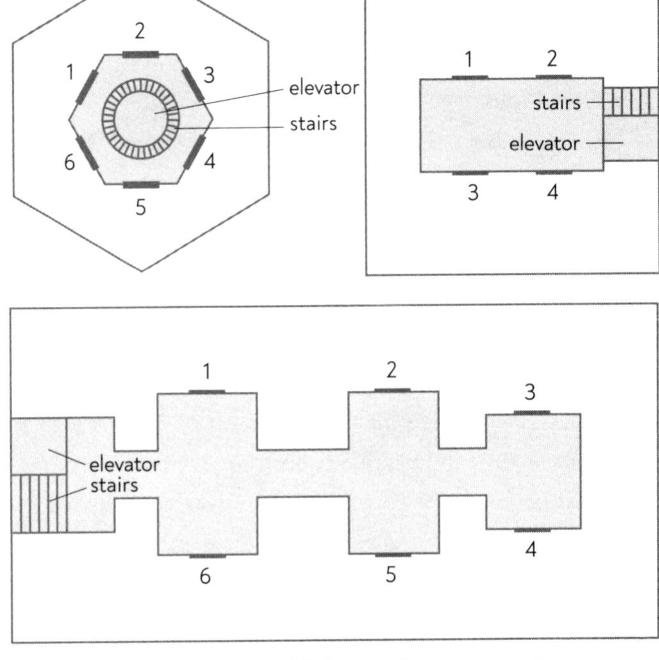

FIGURE 8A. Desirable layout of apartment units.
Not all units suffer from negative energy.

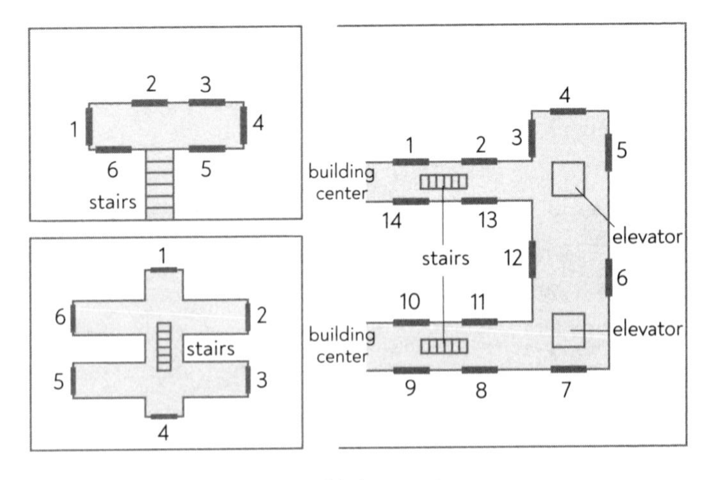

FIGURE 8B. Undesirable layout of apartment units.
All units suffer from some kind of negative energy.

are used constantly, this swirling energy will affect the unit both day and night. Moreover, because the elevator shaft is a deep "hole," units next to the elevator are like houses perched on a cliff edge. No one wants to live next to a deep black hole. Units opposite the elevator will be affected by restless energy although they will not have the effect of living at the edge of a black hole.

Stairs have effects similar to elevators. The only difference is that traffic along stairways is slower. Hence, the upward and downward rush of energy is less.

Architecture of Surrounding Buildings

The energy of a building is also affected by the architecture of the buildings around it.

Relative Size of Surrounding Buildings

Low-level buildings situated along streets with slow traffic that are also surrounded by equally low structures are desirable to live and work in. Energy can flow unhindered, and no building will gather more energy than another.

If there is an uneven distribution of low and high buildings, energy will not flow smoothly. Because tall buildings tend to absorb more energy than shorter buildings, the smaller structures will receive less energy.

If a street is lined by tall buildings, the height of the buildings on either side of the street will form an urban "canyon." The energy along these streets tends to be squeezed and will have little chance to flow inside the buildings.

The worst situation is one in which a small low-level building is totally surrounded by tall buildings. This building will be dominated and intimidated by its tall neighbors. Moreover, because the larger buildings will gather the lion's share of any positive energy, there will be little to none left for the small building.

The Look and Feel of Surrounding Buildings

As mentioned earlier, buildings carry energy through their "look and feel." The appearance of a building gives it a "personality," whether it is positive and nourishing, or negative and destructive.

Architectural designs that carry benevolent energy are described on page 67. If you have a view of these types of architecture, you will receive positive energy from them. Architectural designs that carry aggressive or unstable energy are described on page 64 and 65. If you have a view of these types of architecture, you will receive negative energy from them. You will get the most impact, both positive and negative, if you are immediately adjacent to the building. The impact is lessened if you are not immediately adjacent but still have a full and clear view of the building's architectural features. Some small ambient impact can be experienced if you live near but do not have a clear view of the architectural features.

Landscaping around the Building

Energy entering a building is affected by landscape features surrounding it. Of foremost importance is energy coming in from the front. In fengshui this is called the "welcoming" energy, regardless of whether the energy entering is positive or negative. If it is a single house or a duplex, the house will be impacted fully by the landscape in front. If it is an apartment complex, all the floors will be affected, with lower levels being impacted most.

Second in importance is the energy coming from the back. Typically, a building is less protected at the back. This means that although the energy has less impact, it can hit the building where it is more vulnerable.

Energy coming from the side has the least impact. Typically, the size of most side yards is small and therefore there is less chance for harmful objects or landscape to be situated there.

Landscaping in your neighbor's property will also affect your building energetically. The greatest impact comes from landscaping that runs directly across the back of your building. Next in impact comes from landscaping to the side of your building. The least impact is from landscaping across the street.

Refer to chapter 11, "Object Energy," on page 88 to evaluate the effects of landscape features surrounding a home.

Outdoor Architectural Features

Outdoor architectural features can affect the fengshui of a home. These include driveways, bridges, verandas, gazebos, decks, pools, patios, and outdoor kitchens.

Evaluating the Fengshui of Outdoor Architectural Features

- For a single home, the backyard should be larger than the front yard. The extra space toward the back of the home offers much needed protection for rooms, typically the bedrooms that are located at the back of the house.
- The driveway should not run straight toward the living areas of the house. Driveways running directly into living rooms, kitchen, family room, and bedrooms will bring destructive energy into those areas. It is as if the vehicles are about to crash into people sleeping, eating, or relaxing there.
- Walls or fencing surrounding a property should not carry sharp and pointed features because it introduces an energy of aggression around the home.
- For the same reason, gates should also not have aggressive-looking features.
- The entrance pathway toward the main entrance should not be long, dark, and narrow. Dark and hidden unpredictable energy will ambush you as you enter or leave the home.
- Fountains can be in front or at the back of the

house. However, make sure that the water from the fountain does not flow away from the house. Water flowing away from a home means wealth is being drained from the home.

- Gazebos, outdoor kitchens, ponds, pools, and hot tubs should not dwarf the house. If these features are too large relative to the house, they become intimidating rather than beneficial.
- Trellises and pergolas are desirable because they create a buffer between the house and the outside world. However, make sure that these features do not have sharp, pointed features.
- Bridges or covered walkways should not connect two sections of a house. These structures can isolate areas of the home energetically. This can have a negative impact on communication between occupants of those rooms. For example, you should not have a long walkway separating the master bedroom and the children's bedrooms.

  There are a few exceptions where a separation between two areas of a house is desirable. For example, a covered walkway between a house and a separate garage or between a house and a professional home office or between a house and an income-generating rental unit.
- Verandas, balconies, decks, and outside stairways should be supported by strong thick pillars. This ensures that the structures have an energetically strong base.

11

Object Energy

THE FENGSHUI OF THINGS

OBJECTS CARRY ENERGY, whether they are living things like plants or human-made ornaments and structures. The energy of an object is carried in its shape and appearance. Thus, sharp objects, like iron spikes, glass shards, cactus spines, and sculptures resembling spears and arrows, carry penetrating destructive energy. Objects that are pleasant to look at, like flowers, elegant ornaments with smooth surfaces, and clear ponds, carry benevolent energy. Objects that carry benevolent energy have good fengshui, and objects that carry destructive energy have bad fengshui.

Object energy is about how you feel an object is affecting you. No one wants to wake up in the morning to a view of garbage dumpsters. On the contrary, it feels good to dine in front of a beautiful garden.

It is therefore important that you look around outside a home and identify objects with good and bad fengshui before you decide to move in. If you are currently living there, you may want to introduce objects with good fengshui and use countermeasures to ward off objects with bad fengshui. The use of countermeasures against adverse objects is discussed in part three.

Outdoor Landscape Features

Landscape and Human-Made Garden Features That Carry Positive Energy

Flower beds, vegetable and herb gardens, small bushes with broad leaves, and healthy-looking trees all carry positive energy. Bird feeders and baths that are elegant and do not have aggressive structures such as sharp edges and points also carry positive energy. An added bonus is if birds are attracted to them. Animals bring life energy, whether they are pets or wildlife. Play equipment and tree houses that are well-maintained and do not dominate the backyard also carry positive energy, especially when they are used frequently by children. Young children carry the energy of growth, and play activity carries vibrant and lively energy. Gazebos and lawn furniture that do not have sharp edges and sharp points also carry positive energy.

Waterfalls can introduce the energy of wealth and elegance into an area, but the water must flow toward the home. Water flowing away from the home will carry away the energy of health and wealth. Ponds introduce water energy that is associated with wealth. However, there must be movement in the water. It is better to have no pond than to have a stagnant pond with dirty water. Swimming pools and hot tubs also carry water energy, but their size should not dwarf the home or take up most of the area of the outdoor space. Otherwise, they will take the lion's share of energy away from the home.

Landscape and Human-Made Garden Features That Carry Negative Energy

Plants with aggressive-looking shapes, cacti with large spines, succulents with blade-like leaves, trees with gnarly branches, and roots that resemble coiled snakes all carry negative energy. Broken-down sheds, abandoned play equipment, rusting equipment and vehicles, cracked patio pavers, and weed-infested gardens all carry decaying energy. Structures that have sharp edges and points carry aggressive energy that can "attack" the home as well as send aggressive and disharmonious energies toward neighbors. Garden sculptures that resemble weapons will bring conflict and aggressive and destructive energy into the area. Most destructive are actual weapons that are displayed as sculptures. I once saw an old tank displayed in a war memorial park, with its guns pointing at a row of houses across from it. Old forts that have been transformed into museums often keep cannons displayed on the walls. In one town, I saw cannons pointing at city hall! Needless to say, you don't want to live or work in view of weapons pointing at you, even when they are clearly museum pieces.

Note also that equipment and rusting vehicles that are used as sculptures still carry negative energy. A broken tractor in the driveway is still a broken tractor regardless of whether it is treated as a sculpture or junk.

Utility-Related Structures

Utility-related structures include electrical transformers, power lines, satellite dishes, antennas, and cell phone towers. These structures carry volatile energy that can disrupt human and natural energy. Their features are typically sharp and pointed, and therefore introduce aggressive energy into the area.

Buildings situated close to these structures will suffer from negative energy that affects health, livelihood, and harmony. Most affected are homes that have a direct view of these structures.

Depending on the size of the utility structure, the effect can be large or small. For example, a power transfer station with many high-tension wires will impact a large area, including homes that do not have a direct view of them. Electrical transformers and power lines along a street will affect the homes adjacent to them and those that have a view of them. Satellite dishes are gigantic reflectors that bounce signals toward homes located around them. These signals can also disrupt health and livelihood energy.

Generally, you can think of utility-related structures as creators of tornado-like vortices of energy, sweeping across neighborhoods that are in their radius of influence. If you are living adjacent to large utility-related structures, the best option is to relocate. If relocation is not viable, there are countermeasures that you can install to mitigate some of the negative effects. You can find these countermeasures described in part three of this book. However, countermeasures can only minimize the negative effect. They cannot remove the effect altogether.

12

Moving Inside

ONCE YOU HAVE DETERMINED that there are no negative energies surrounding a building, you can begin to evaluate the fengshui of the inside. Follow these steps when you are evaluating the fengshui of the interior of a home.

1. Examine the floor plan.
2. Walk through the space.
3. Examine the interior architectural and design features of the home.
4. See if the space fulfills your needs.

It is important you follow these steps sequentially. If you get attracted to unimportant features you may neglect the fengshui that can affect your health and livelihood. Often, real estate agents try to persuade clients to sign a contract by drawing their attention to a beautiful shower stall or the tile on the kitchen floor. There is even a cliché among real estate agents: "Bathrooms and kitchens sell a home."

Step 1: Examine the Floor Plan

Before you walk through a building, you should look at its floor plan. First, the floor plan gives you a bird's eye view of the layout of the home that you won't get by walking through it. Second, the floor plan allows you to see the layout abstractly without being distracted by the architectural features, the decor, and the furniture. Third, you can save a lot of time in your search without having to make appointments to physically walk through a building. If you learn how to do an initial evaluation of the fengshui of a home by examining floor plans, you will be able to narrow your search and only visit buildings that pass your initial scrutiny.

Let's look at the important aspects of the floor plan.

Entrance

The main entrance should not line up with large windows or doors at the back of the home. Any benevolent energy entering the main entrance will go straight out through the back.

Stairs

If the stairs are aligned directly with the main entrance,

domestic energy from upstairs will leave the home. If there is a home office upstairs, the wealth will flow out of the home as well. Additionally, untamed energy entering the home from outside will flow upstairs to the sleeping or work areas unchecked. In split-level homes, if there are stairs flowing directly downward from the main entrance to a lower level, any benevolent energy flowing into the home from the main entrance will get sucked into the lower level. As a result, the upper levels will not be able to benefit from the positive energy entering the building.

Windows

There should not be too many floor-to-ceiling windows. Large windows allow harmful energy to enter and beneficial energy to leak out.

Doors

Doors of bedrooms and home offices should not face stairways that exit the building. If the doors of bedrooms and home offices open directly toward a stairway, this means that energy that benefits activities in those rooms will flow out and not get accumulated.

Corridors and Hallways

Corridors should not be maze-like. Energy flowing through a maze will be trapped and transformed into confused and anxious energy. In addition, positive energy will be dissipated by the time it reaches rooms located at the end of the maze.

Hallways should not be long and narrow. Energy flowing through narrow hallways is squeezed and choked, and rooms opening into the hallways will not get sufficient energy entering them. Moreover, energy flowing through a long narrow space will be squeezed and transformed into frenetic and fearful energy. As a result, rooms opening into the narrow hallway will receive energy that has been transformed negatively by the narrow space.

Kitchen

The kitchen should not be located near the main entrance, and should definitely not be the space you walk into when you enter the home. Kitchen energy is associated with livelihood. In Chinese culture, we call it "having enough to eat." If the kitchen is the first space you enter into from the main entrance, livelihood energy will flow out and will not be able to gather. The kitchen is also associated with family and domestic energy. If the kitchen is the first space you encounter when you enter a home, family energy will be restless and unstable. Because the entrance is the area where the home meets the world, the energy here tends to be unstable, untamed, and restless. Unstable energy needs a resting area so that it can be calmed and slowed before it reaches the living areas of the home. And this place should *not* be the kitchen.

Living Room and Family Room

The formal living room should be located toward the front of a home. This allows visitors to gather without having to walk through the family's private spaces. You do not want your guests to walk past your bedrooms toward where they will be entertained.

If you want an informal family gathering area in addition to a living room, this space should be away from the main entrance of the building and be open toward the kitchen. Both are "domestic" areas. If they are connected, the domestic energy and harmony of the family will be strengthened.

Dining Room

For most families, the formal dining room is rarely used for dining except during special holidays or celebrations. Most people dine in a smaller eating area close to and open to the kitchen. Dining areas, whether formal or informal, should be located in the middle of the home and away from the main entrance. In this way, disruptive energy entering from the outside will not reach the dining area immediately. A good and relaxed meal is important for replenishing energy. Having disruptive energy buffeting the dining area will introduce disharmonious and possibly quarrelsome energy while the family is having meals. Most importantly, dining rooms should not be "passage" ways between spaces. No one wants to dine in an area that is a busy thoroughfare.

Bedrooms

Bedrooms are where we replenish our energy with restful sleep. Therefore they should be in the most protected part of the building. If there is an upper level to the house, the bedrooms should be located upstairs. If they are in the main level of a house, they should be located toward the back of the home. In this way, the bedroom will be protected from unpredictable and untamed energy entering through the front door. In apartments, they should be located as far away as possible from the entrance to the unit.

Bedrooms should never be on top of or adjacent to the garage. Vehicles entering and leaving the garage can bring restless energy into these areas. Moreover, any unpredictable and possibly negative energy picked up by the vehicles on the road will be carried into the bedroom.

Home Office

While any room can be used as a home office, the best location for a home office is a room with a view. Having a view means that business conducted in that room will have vision and innovation. Never put a home office in a basement.

Bathrooms

Bathrooms should be located conveniently and should not be too small or too large. They are not considered "important" rooms because we don't spend a lot of time in them, compared to bedrooms, home offices, kitchens, and family rooms. In general, bathrooms should not be larger than a bedroom or a home office.

Relative Size of Rooms

Compare the relative sizes of the rooms. Important activities should occur in rooms that are largest. A large space allows more energy to gather, giving the most frequently used rooms the biggest share of overall energy in the home. A room that you do not spend much time in should not be larger than the ones you do spend time in. Closets, bathrooms, and storage rooms should therefore be smaller than bedrooms, home offices, and family gathering spaces. If the en suite bathroom is larger than the bedroom, or the closet is larger than the home office, then there will be a confusion of priorities. For example, if the clothing closet is larger than the home office, you will likely spend more energy deciding what to wear than figuring out how to improve your finances. If your home office is massively larger than your bedroom, your business concerns will overshadow your domestic life.

Garage

The entrance to the garage should not "lead" directly toward the living space. In other words, vehicles entering a garage should not be "driving" straight at kitchens, bedrooms, or home offices. The best location for a garage is in a space unconnected to any living space. The next best option is a location that is parallel to the building, so that vehicles entering the garage do not "drive into" any living space. If this is not possible, then it would be desirable to have a mud room, laundry/utility room, or storage space against the garage to buffer the garage space from the living space.

Bedrooms, kitchens, and home offices should not be located directly above the garage. Garages are where vehicles enter and exit. The movement of vehicles underneath a space makes energy above it unstable and restless. A bedroom above a garage will not allow the occupants to have a restful sleep. Home offices above a garage will mean instability in business, and kitchens above a garage will mean instability in livelihood and domestic harmony.

Basements

Basements are common in North American houses and are often made into living spaces. Traditionally, basements were called cellars and were built as storage space; they were not meant to be lived in. In old houses, basements are often heated less efficiently and are dark and

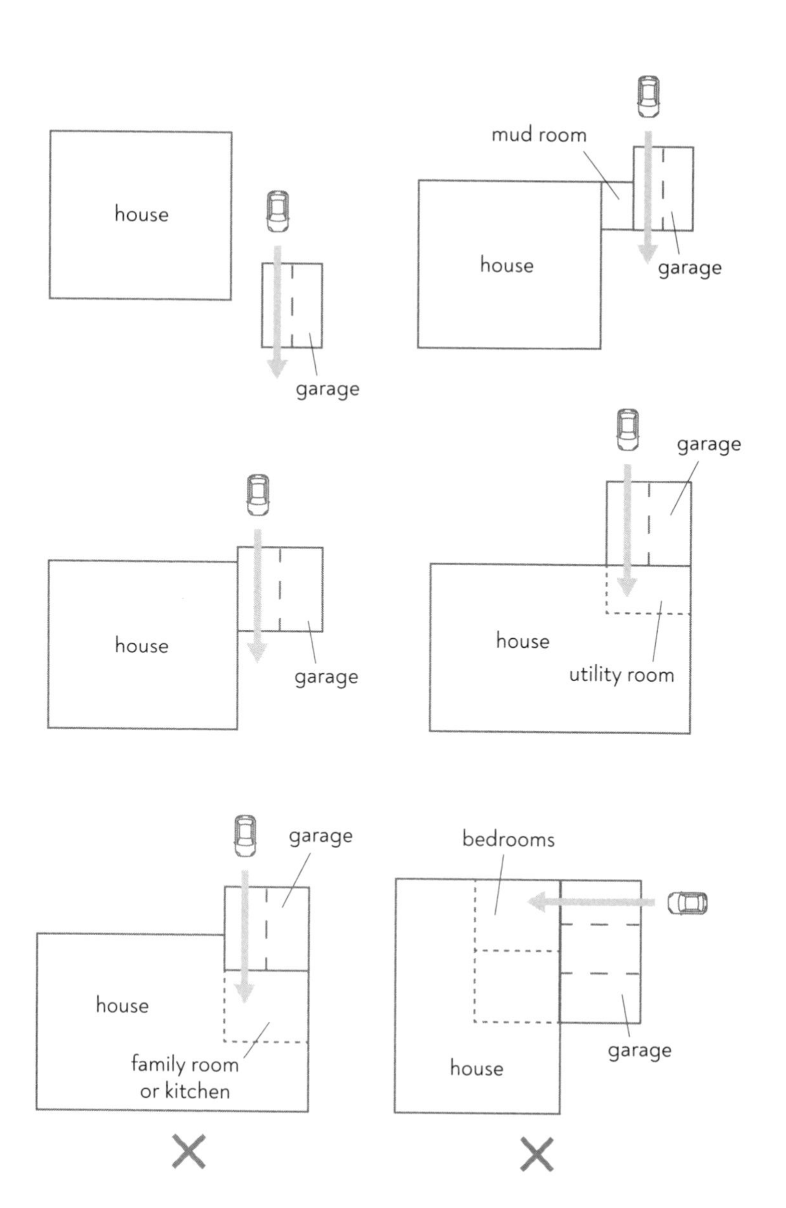

FIGURE 9. Desirable and undesirable garage locations.

damp. Even in modern homes, heated basements are often colder than the main and upper levels of the building.

Guidelines for Considering the Use of Basement Spaces

- Basements are accessed by stairs leading downward. Positive energy tends to rise and negative energy tends to fall. This means that positive energy entering a building will have more difficulty getting to the basement than to the upper levels. Life energy and wealth energy in basements will therefore be less. Also, more negative energy will tend to gather in the basement. This problem is exacerbated if windows in the basement are small and the ceilings are low. All in all, even finished and heated basements are not ideal areas to locate bedrooms and home offices. Home offices need views to the outside world in order to project and receive worldly energy. Bedrooms are spaces where we gather energy when we sleep. If there is insufficient energy, we will not be able to get the maximum benefit from sleep as a way of replenishing energy. Most importantly, because of limited access, basements' energy tends to be isolated and does not mix well with energy in the rest of a house. People who live in basements tend to spend most of their time there, away from those who live in the upper levels of the house. It is for this reason that children should not have bedrooms in a basement.

- Basements are best used as storage areas or for activities that are not related to wealth and rest. You can use basements for recreation and entertainment. It is also viable to place guest rooms in the basement since hopefully guests will not be staying there for more than a month. Usually after a month, the energy of the space will start to affect the health, the livelihood, and the well-being of the individual living there.
- If the basement is a "walk-out" basement, its energy is different from a true basement. A walk-out basement is a space that has one or more sides open to the outside. Large windows and glass doors allow energy to enter, making the area more like a main level of a house than a basement. The walk-out basement can be used in the same way as the upper level of the house. There are two things you need to note about walk-out basements: One, if there is wall separating the part of the basement with external access, the part of the space with no windows is considered a true basement. Two, a basement is not considered as a walk-out basement if the only external access is a solid door leading outside.
- Basements underneath a garage should not even be used for human activity. Cars drive over it and you will feel you are run over by vehicles all the time. Areas of basements underneath a garage should be used strictly for storage.

- If you are thinking about finishing a basement, make sure you plan on how to use the space first. Remember that positive energy is depleted when it enters a basement. Therefore make sure that activities in the basement do not isolate family members from each other. A family spending time together harmoniously is a happy family.

Outer, Inner, and Private Zones of Living Space

Homes can be divided into three zones: outer, inner, and private.

The outer zone is typically the area of the home closest to the entrance. It is the zone where visitors first encounter your home. Ideally, the entrance area (or vestibule), the living room, the formal dining room, and a small powder room should be located here. Sometimes this is called the public zone because it is where the public, or persons not close to the occupants, are welcomed. It is also where visitors would normally gather. The outer zone is "open" to anyone who visits—relatives, close friends, acquaintances, and even strangers. If you are throwing a neighborhood party or a holiday dinner for your colleagues from work, you would prefer your guests to stay in this area.

The inner zone is the area of the home reserved for the family and close relatives and friends. The kitchen, family gathering space, and the informal eating area are considered part of the inner zone. In most homes, the inner zone is situated in the middle of the building. If the home opens to a backyard, the inner zone can lead to the backyard where family and friends can gather casually. The inner zone is not meant to be frequented regularly by causal visitors and strangers. It is an area where domestic energy gathers, and having too many casual visitors frequenting this space will disperse or disrupt domestic energy gathered by the family and those close to the family.

The private zone is the area of the home that is restricted to only the occupants. In this zone are the home office, bedrooms, the walk-in closets associated with the bedrooms, and the en suite baths. Ideally, these rooms should not be open to friends or even visiting relatives. Bedrooms carry and accumulate personal energy of those who sleep there. Introducing energy from visitors can disrupt and dissipate the personal energy gathered by the occupant. It is not just "turf" space that cannot be intruded into. When siblings do not allow other siblings and their guests into their bedrooms, there is an intuitive reason behind it. It is likely that the occupant feels that their energy is being intruded into. This is the same reason that children need to move out of the parents' room when they are no longer in a crib. Even toddlers need to have their own space for their energy to gather.

The home office is considered a private zone because financial activities are considered private. Your bank accounts, investment portfolios, business plans, and personal and business accounting need to be protected, and more importantly, business energy needs to be contained and not leak out.

Needless to say, private zones should be off-limits to visitors. Ideally, private zones should be located in areas that are the least accessible and the farthest away from the main entrance. If a home has two levels, then bedrooms and the home office should be located on the upper level. If there is not enough space for a home office on the upper floor, then the home office can be located at one side of the building away from the living room and kitchen area. Never "carve" an area of the kitchen into a home office. First, it is too exposed. Second, with family traffic moving around the area, business planning and activity will be disrupted, and no wealth can be accumulated.

If your home office is used as a professional space that receives clients, ideally that space should have a separate entrance so that your business guests do not have to walk through the home to reach the office. In a nutshell, you do not want public traffic moving through the private zone of the home to do business with you.

Navigating guests through your home is easier than you think. Just make some simple rules:

- Don't invite guests and strangers into your bedrooms, not even to use the en suite bath.
- Don't invite guests and strangers into your home office even if they ask if they can use your computer to check messages.
- If you throw a big party that spills into the backyard, try to move the crowd to the yard or

patio as smoothly and quickly as possible. Don't encourage people to stick around in the kitchen.

- Close bedroom and home office doors when you expect unfamiliar visitors or strangers. Most people will not push open closed doors uninvited. However, an open door is an invitation for anyone to enter.

Having clear outer, inner, and private zones will help you organize activities, minimize confusion, and improve harmony among family members and housemates. Therefore, when you are looking to purchase or lease a building, you should get a feel for whether the floor plan allows you to have a clear distinction of the three zones: outer, inner, and private.

Additional Considerations for Apartments

Generally, the rules for choosing an apartment are the same as those for single homes and duplexes. There are a few distinct considerations, however, that you need to pay attention to in the layout of units in the apartment building.

First, examine the locations of elevators and stairways. Units near or facing elevators and stairways are not desirable. This was covered earlier.

Second, weigh the trade-off between views and protection. Corner apartments have more views, but because they are more exposed, it is also easier for both positive and negative energy to enter. If there are destructive structures such as aggressive architecture or transmission towers close by, having corner views may expose you more to these negative energies.

Third, basement apartments should be avoided regardless of whether they are part of a large apartment complex or a unit attached to a single home. Positive energy cannot enter basement units easily, not to mention there is a higher risk of flooding during storms.

Step 2: Walk through the Space

If the floor plan of a space has no jarring problems, the next step is to arrange a visit and walk through the space. Walking through a space physically allows you to tune in to the details of how the space is configured, how each room is laid out, and what the interior built-in architectural features look and feel like.

Never sign a purchase or rental contract based on a virtual walk-through. You need to be physically there to get the feel of the energy inside the space. And bring a tape measure with you so that you can plan how to use a space effectively and see what type and size of furniture will fit a room.

Here is a list of what you should do when you walk through a space:

- **Examine the most important rooms first:** Rooms important to you are the bedrooms, kitchen, home office, and the family room. If the formal dining room is used as the main eating area, then it is considered an important room. Secondary rooms are recreation rooms, bathrooms, utility rooms, and storage rooms.

- **Walk from one end of the space to the other:** If you can do it without making a lot of turns, this tells you that the space is not maze-like. The minimum width of a hallway regulated by building code is three feet (approximately one meter). Any width less than this measurement is considered too narrow. Preferably, a hallway should be wider than the minimum code. If the length of a hallway is longer than twelve natural paces *and* if it is barely three feet wide, the space is considered long and narrow and therefore problematic.
- **Walk from one room to another:** If you can walk from one room to another easily without having to meander around, this also tells you that energy can flow easily from one room to another.

Bedrooms

Bedrooms should have windows that let in sufficient light. Light energy is life energy. However, floor-to-ceiling windows are considered problematic. Not only do they offer no protection, but they allow life energy to escape. Look outside the bedroom windows to see if there are negative or destructive features. What you can see will "enter" through the window and affect you.

As mentioned earlier, the bedrooms of residents should not be located in a basement. Life energy cannot get into basements easily, and negative energy tends to gather and get stuck there. Most importantly, if some members of a family have bedrooms in the basement, they will be isolated from other family members. Guest bedrooms are viable in the basement because guests are not regular occupants and will stay only for a limited time.

Kitchen

The kitchen is where the energies of health and livelihood are gathered. Move around the kitchen as if you would use it. Don't get distracted by the shiny new appliances or tiles on the backsplash or floor. Make sure that when you cook, your back is not exposed to the main entrance. You should be able to see who's entering the kitchen area when you prepare food. Most importantly, confirm that the kitchen is not the immediate space you walk into when you enter the home. Finally, you should make sure you are not standing under a beam when you cook or when you are eating at the kitchen countertop.

Home Office

If you are planning to have a home office where you will be engaging in money-making activity, make sure the office can be located in a room with a good view to the outside. You should also look outside the windows of a home office and check to see if there are any negative or destructive features in view. The energy of these kinds of features can obstruct and even sabotage your business activities. Business vision is related to how far you can see out of the home office windows. For this reason, home offices should not be located in basements.

Family Gathering Room

The family room is associated with the energy of harmony and mutual support among family members. Make sure that there is a space where the family can gather and spend quality time together.

Ceilings and Exposed Beams

Make sure you examine the structure of the ceilings in the rooms. Flat ceilings are safest because energy can circulate easily and not get trapped in constricted areas. Vaulted and trapezoid ceilings are viable if they are higher than nine feet (approximately three meters). Next, check for the presence of beams. Exposed beams, whether structural or faux, have strong energetic effects on a space. Beams are associated with the energy of heavy burdens. Living under ceilings dominated by beams will make you feel like you are carrying a heavy weight. The higher the beams are above the ground, the less effect it will have on the occupants. If you want beams in your home, ideally they should be on ceilings at least fifteen feet (approximately five meters) high. Even so, you should not position beds, couches, dining tables, or workstations directly under the beams.

Beams over sleeping areas are especially harmful. If the beam lies across the bed horizontally, it will have a negative effect on the health of those sleeping under it. For example, if the beam is across the legs, the occupants may have leg-related problems. If the beam is across the stomach, the occupants may have bowel-related problems. If the ceiling beam lies lengthwise down the bed, separating the bed into two halves, this could result in relationship problems.

One final bit of advice: Don't get distracted by interior designs. Evaluate the fengshui first, and then consider whether you like the interiors. If you are too enamored of the material on the countertops, you may neglect to pay attention to where the cooking stove is located. If you are too excited by the tile on the bathroom floor or the shower stall, you may neglect to pay attention to whether you can see aggressive objects from the bedroom window.

When you walk through a prospective property for purchase or rental, try to do it without the realtor tagging close behind. Many agents will try to sell you the property by pointing out how attractive the tile is, how good the hardwood floors look, and so on. Walk through the building slowly and quietly. Go through the checklist above and evaluate each space carefully and thoroughly.

If you are evaluating the fengshui of a space you are now living in, the same principles apply. However, if you see problems, you will need to install countermeasures to mitigate them. These methods of mitigation can be found in part 3 of this book, "Fengshui Strategies."

Step 3: Examine the Interior Architectural and Design Features

Interior architectural features include structures like fireplaces, wall features, decorative beams, light fixtures, ceiling fans, and built-in furniture units. Interior design features include wall coverings, such as wallpaper and paint, as well as tile, kitchen and bath fixtures, and flooring.

Fireplaces

Fireplaces are meant to heat a space. However, in modern living, the fireplace is also a design feature that can add elegance and beauty to a room. Regardless of their function, fireplaces should not dominate a space. If the size of the fireplace is too large, fire energy will be too strong. Fire is a volatile and domineering element, and a strong presence of fire can lead to emotional flare-ups and arguments. The feature wall above the fireplace also affects the energy of the room; it can intensify the energy of fire or calm it down. The best type of wall feature above a fireplace is a plain wall painted with a neutral color. The next best type of wall feature is a tile or smooth stone slab that has neutral color such as gray or white. The worst type is a feature made of rough-hewn rocks. The jagged rocks add aggressive energy to an already aggressive element such as fire. Needless to say, over-large fireplaces and jagged rock walls above them are not desirable for living rooms, bedrooms, or family gathering space. Fireplaces are difficult and costly to remove or restructure. Therefore, if the fireplace is problematic, you should think about whether you can redesign it easily before you decide to purchase or rent the space.

Light Fixtures

Light fixtures hang from ceilings. As such their energy "hovers" over the room. Their effects are greatest on areas that they are directly over. The best type of light fixtures are those that hug the ceiling and do not hang excessively low. Low-hanging light fixtures are like obstacles that threaten to fall on the residents, especially if they are hanging from low ceilings. Other problematic types of light fixtures are those that have sharp and pointed features. These types of fixtures act like spears pointing down on anyone who walks, sits, or sleeps under them. The same principle applies to ceiling fans. Ceiling fans have whirling blades, and it is not desirable to sleep or work directly under the blades, especially if they hang from low ceilings.

Light fixtures and ceiling fans are not difficult to change. If they are problematic, you can replace them. There are now ceiling fans that have small "blades" hidden inside a cage that render the blades invisible. There are also ceiling fans with blades that look like leaves and fronds. In fengshui, if a feature is not seen, it will not have an effect. If an aggressive feature is transformed into a harmless or even elegant feature, its energy is changed.

Built-In Furniture

Built-in furniture units also carry energy that can affect a space. The general rule of thumb for evaluating built-in units is that they should not be the dominant feature in a room. Ideally, they should occupy at most 30 percent of the space. This applies to built-in desks, workstations, and bookshelves.

Wallpaper

Wallpaper patterns should not be too dark or busy. Dark colors bring heavy energy that absorbs light and life energy, leaving none for the occupants in the space. Busy patterns can distract you from activity you do in that room. Many people find busy wallpaper patterns overwhelming. You may find some busy patterns whimsical or attractive on initial observation, but think about what it would be like to live with them all the time. Fortunately, it is not difficult to change wallpaper or remove it altogether. Therefore, wallpaper that appears too dark or too busy should not be a reason for you to reject a property.

Paint Colors

The paint colors on the walls should not be too dark or busy either. The safest color for walls is a neutral or pastel color. Dark colors bring intense energy. The bottom line is this: Walls painted a light color mean a life of light and simplicity. Walls painted a dark color mean a life filled with heaviness and lethargy. Fortunately, like wallpaper, paint colors on walls can be changed easily.

Tile Patterns

Tile patterns can also affect the energy of space. Depending on how much space the tiles cover, the effects will vary. For example, a wall completely covered by tiles has a bigger effect on a space than a small backsplash on a kitchen wall. Tile patterns that are simple and not busy are the best. Complex patterns generate unnecessary complexities in the activities in that area. Busy tile patterns can also be a hazard. Fallen objects on kitchen floors can be camouflaged by a busy floor pattern and become a trip hazard. A busy tile pattern on a bathroom floor is especially dangerous since we are generally barefooted in that room and can step on sharp objects that are camouflaged by a busy tile pattern. Because tiling on floors and walls can be changed with moderate cost and effort, you should not let a problematic tile pattern deter you from purchasing a home.

Countertops

Countertops should not have a busy pattern either. Newer homes tend to have quartzite, granite, or marble countertops. Regardless of the type of material, patterns in the stone should be uniform and not busy. A busy stone pattern on a countertop will camouflage food and utensils. I know people who thought they had lost knives and later found them on the countertop. A surface with large fissure-like pattern carries the energy of cracking and breaking even though it is a hard stone. In a kitchen, the energy of breaking and cracking is associated with the breakage of dishes, food spilling, and most dangerously, the breaking of interpersonal relationships associated with kitchen activity, be it relationships among family members or between friends invited to a social gathering.

While it is relatively easy to change fireplaces, light fixtures, ceiling fans, floor materials, backsplashes, and countertops in an owned home, rental properties are more restrictive. Choosing a rental property is therefore more challenging. Thanks to internet technology, though, you can look at rental properties online and weed out rentals with problematic interior architecture or design.

Decorative Colors

Colors have energy. Their energetic essence is based on the cosmology of the five elements in Chinese culture, which was mentioned in chapter 10 (page 72). The theory of the five elements is ancient; it is said to have originated several thousand years ago. This cosmology of classifying all objects into five categories has for centuries influenced Chinese medicine, the healing arts, divination, and of course fengshui.

Each of the five elements—metal, wood, water, fire, and earth—is associated with a color. Metal is manifested as silver or gold; wood as green; water as blue; fire as red; and earth as brown and yellow.

Guidelines for Using Colors in a Home

- **Silver and gold can be used anywhere:** While they are neutral in spaces like the kitchen or family rooms, they are best used in home offices because gold and silver are associated with the energy of wealth.
- **Green is a soothing color:** It is the color of trees, leaves, and growth. It brings nourishing energy into a space, and therefore, is most suitable in bedrooms.
- **Blue is associated with the element of water, which is soft and nourishing:** This color can dissipate intense energy and is best used in rooms where potential conflicts can occur, such as the kitchen and family rooms.
- **Red is the color of fire:** Fire is a dangerous and volatile element. If a kitchen is dominated by the color red, there will be burn and fire hazards. If

red is dominant in a bedroom, there will be frequent emotional outbursts and quarrels between partners. If red is dominant above a fireplace, there will be risks of fires involving the chimney and the hearth.

- **Yellow and brown are colors that carry the energy of earth:** These are neutral colors and can be used anywhere.

Repainting walls and moving decorative objects around is not difficult. Even if you are renting a property, most owners will let you change the color of the walls. Furniture upholstery, bed covers, and cushions can be covered to minimize negative energy associated with colors. Art, wall hangings, and window covers can be chosen to avoid introducing negative energy into a room.

You can increase the energy of the color green by placing plants in the room as well as having a highlight wall painted in a greenish color. You can increase the energy of the color blue by placing art and decorative objects that are blue in color around a room. You can introduce gold or silver into a home office with metal furniture, lamps, and storage units that have a silver tint. You can avoid volatile red energy by removing red coverings on furniture in the bedrooms and on couches near a fireplace. Finally, you can introduce earth-toned colors with ceramics.

Is the color red totally unusable? Not if you get a professional fengshui consultant to help you to use the color red strategically. Some areas of a room can be enhanced by the color red. However it takes someone who is skilled in the school of Flying Stars fengshui to identify areas that can be enhanced by the color red. If you are doing your own fengshui evaluation, it is best to *play it safe* and stay away from using red.

Step 4: Plan Furniture Placement

Bedrooms

Bedrooms are the most important rooms in a home because we spend most of our time there. Bedroom energy is typically associated with health, relationships, and nourishment. We gather and replenish energy through sleep, and since most people sleep seven to eight hours a day, we end up spending a good deal of our time in the bedroom. Moreover, intimate relationships with our partners typically occur in the bedroom, and many people like to read or watch night television in the bedroom before going to sleep. All in all, the time we spend in the bedroom dwarfs our time in any other room in the home.

The most important furniture in a bedroom is of course the bed. Beds are where we sleep, rest, and have intimate relationships with our partners. Parents also play or read to young children in bed.

Guidelines for Placing a Bed

- Don't place the head of the bed against a window. Windows let in unpredictable energy from the outside. Areas against windows are the least protected section of a room.

- Don't place the side of the bed against the entrance door to the room. Unpredictable energy entering the room will run into the bed first.
- Don't place the bed in front a mirror. You should not be able to see your image in the mirror when you are sitting or lying on the bed. Mirrors create a second image of a person. Life energy will be split between the real person and the "reflected person. As a result, you will receive only one-half of the life energy available in the room.
- The placement of a crib follows the same principles as the placement of a bed.
- Nightstands should be scaled appropriately in relationship to the bedroom. Don't fill the surface area of the nightstand with many objects. A crowded nightstand introduces "busy" energy into an area where you should be sleeping and resting.
- Do not crowd the bedroom with large pieces of furniture such as free-standing storage units, dressers, desks, and so on. Furniture occupies space, and the less open space you have in a bedroom, the less life energy can gather and circulate in the room. A general rule of thumb for making sure that the room has enough space for you is to see how freely you can walk around the room without running into furniture.

- If there is a ceiling light, make sure that the ceiling light above the bed does not have sharp features pointing down toward the bed.
- If there is a ceiling fan, make sure that the ceiling fan is not too large and does not hang too low over the bed. Whirling fan blades over the bed generate energy equivalent to having whirling knives over your sleeping area.

Kitchen

Energy in the kitchen is associated with health and livelihood. Good kitchen fengshui means "having enough to eat," which is the Chinese way of saying that we are well-off. A kitchen with good fengshui also means that the occupants will have no illnesses associated with food. In contrast, kitchens with bad fengshui can bring on illnesses associated with food, whether the food is cooked in the kitchen, consumed as takeout, or eaten in a restaurant.

Guidelines for Placing Appliances and Dining Furniture in the Kitchen

- The cooking area of the kitchen should not be exposed on all sides. You don't want to have the kitchen "floating" in the middle of a space.
- Position the cooktop so that when you are cooking, you will not have your back to the entrance of the home.

- Make sure that the appliances are not red in color. Red carries the energy of fire. The energy of fire is already strong in the kitchen, with flames and heating elements coming from cooktops and ovens. Moreover, the kitchen is the area of the home with the most appliances that draw electrical power, another manifestation of fire.
- Floor and backsplash tiling should not be red for the same reason described above. Desirable colors are blue, green, gray, sand, or white.
- If you have an island that is an eating area, make sure the ceiling lights above it do not hang too low. Otherwise, the light fixtures will create an oppressive and heavy energy over health and livelihood.
- The surface pattern of countertops should not be dark or busy.
- Make sure there are no large ceiling beams over the cooking area or the informal dining area that is part of the kitchen.

Dining Room

While many families rarely use the formal dining area, the dining room is still a prominent feature in a home.

Guidelines for Placing Furniture in the Formal Dining Area

- Make sure the ceiling fixture above the dining table does not hang too low. A general rule of thumb is that it should be at least three feet (approximately one meter) above the food.

- The ceiling light above the dining table should not have pointed features that can attack livelihood.
- When choosing the size of the dining table, make sure that after you have placed the chairs, there is ample space to walk around the table.
- Don't let storage furniture, display cabinets, or serving sideboards dominate the dining room space. Large pieces of furniture in the dining area will take up space that is meant for health and domestic energy.

Home Office

Many people now work full- or part-time from home. In many homes, a space that is designated as an office is becoming more common. There are three kinds of home office: the home finance office, the commercial home office, and the professional office.

HOME FINANCE OFFICE

The home finance office is a space where we do household and personal finances. Perhaps we also check messages from the work office after hours or look through an investment portfolio, but generally, this office is not primarily involved with generating wealth. Furniture can be casual, but the desk or workstation needs to be facing a window with a view. Even if you are not generating wealth in this office, you should have a view that inspires rather than bores you. You should not sit with your back to the door. Preferably, you should have your back against a wall. Storage cabinets and shelves should not dominate the space. This room can also be used as a space for relaxing, reading, or studying. It is also a space that can be shared with other members of the family.

COMMERCIAL HOME OFFICE

The commercial home office is a space where commercial activity is carried out and wealth is generated. The feng-shui of this area therefore affects the livelihood and financial well-being of those living in the home. As more corporate professionals work remotely and more people are self-employed, this space has become an important consideration for buyers and renters. If having a commercial home office is a top priority, make sure you can dedicate a room for this purpose.

Guidelines for Selecting a Room for Commercial Use and Arranging the Furniture

- The office should be a separate and dedicated space. Preferably it should be located as far as possible from any common space used by the rest of the household members. If it is too close to the kitchen or family room, business energy generated in the office will be disrupted or dissipated. Never use a corner of a family room or the kitchen as a commercial home office.

- Do not share your commercial home office space with other members of the family or housemates. Sharing the space will dilute and disrupt the energy that is particular to your business. Ideally, this space should not even be frequented by persons not conducting activity associated with the business.
- The desk and/or workstation that is used to plan and conduct business should have a distant view. The farther and the more unobstructed the view, the bigger the business vision.
- Make sure there is ample space for storage cabinets, shelves, and electronic and professional equipment. Do not pile boxes, files, and objects in the room. A disorganized or cramped office space means a disorganized business or a business squeezed with no option to expand.

PROFESSIONAL OFFICE

The professional office is different from the home commercial office. You will not meet with business clients or business partners in person in the commercial home office. The professional office, however, is a space where you intend to meet with business partners or clients. This being the case, it is best to isolate it from the home.

Guidelines for Situating a Professional Office

- The best option is to situate the professional office in a separate structure not attached to the home, for example in a converted garage or a studio built in the backyard. This allows your business guests to enter the office without going through your living space. Having clients walk through the living space will disrupt the domestic energy in the household. This is especially so for psychotherapists who see clients with emotional issues or health professionals who see patients with physical ailments.

- If it is not viable to have a professional office located separately from the home, the next best option is to situate it in an area where business guests can enter through a separate entrance. An example is a converted attached garage with an independent entrance.
- The professional office should "carry" the energy of the business. Your client should be able to see the accomplishments of your business. For example, architects should display photos and blueprints of their construction projects; therapists should display art that is soothing and relaxing. If the business has a logo, it should be visible to the client when they enter the space.
- Make sure the room is not crowded with large pieces of furniture. Space allows a "spacious" interaction between you and your client or business partner.
- The office should not exude a sterile look. It should also not be overdecorated. Neutral colors are best for a professional office.

Family Room

The family room can have furniture casually arranged so that family members can choose whether they want to be intimate or have some distance from each other. Therefore, furniture should include couches as well as single lounge chairs. Coffee tables should be situated where occupants can use them as shared space. While this is a casual space, it should still have ample room for family members or housemates to move around. This is a domestic space, and as such, furniture placed in this space should facilitate harmony, communication, and joint activities among occupants. For this reason, you should not place large pieces of exercise equipment in the family room. Exercise activity tends to be focused and nonsocial. Only one person can be on a treadmill or stair-climber at a time, and these machines tend to be noisy, making them antisocial.

If it is important for you to have exercise equipment in your home, make sure it can be placed in a separate space such as a basement or a dedicated exercise room. Many apartment buildings do not allow occupants to have exercise machines in their homes because the noise travels easily to the neighboring units, especially those above and below. If the activity of indoor exercise is important to you, you should check to see if the apartment complex has an exercise facility, or if there is a health club close by.

Step 5: Decorate Your Space

Most people like to decorate their space. Whether you are buying, renting, or currently living in the home, decorating the space makes it feel like it is your own. In fengshui, decorations are not just for aesthetic elegance; they can also plant your energy in your space.

You can think about decor before or after you move in. If you are evaluating the decor of a space you are currently living in, you will need to go through every room to make sure there are no spaces with an overwhelming dominance of one color, especially red.

Interior decor can include sculptures, paintings, indoor planters, mirrors, lamps, carpets, family photographs, memorabilia, antiques, and interior accessories such as ornaments, table cloths, bed covers, cushions, and so on.

We all have our aesthetic preferences. However, how we choose our decor can affect the fengshui of a space.

Guidelines on How to Decorate a Space to Optimize Good Fengshui

- Less is better. Objects occupy space. The more objects in a room, the less open space you will have in that room. The less open space, the less room there will be for energy to gather and circulate. This means that the occupants will have less energy to draw from.
- In general, some colors are "safer" than others. The "safe" colors are green, blue, gray, and light

earth tones. Red is a powerful color because it carries the energy of the element fire, a volatile force. It is also an amplifier. This means that the color red will increase the power of energy gathered in the area where it is present. The color red will amplify both positive *and* negative energy. Therefore, it is safer to stay away from using too much red in decor except when a professional fengshui expert can show you how to use red to enhance only the areas with positive energy.

- Make sure you do not put objects with aggressive features in the home. This includes sculptures, memorabilia, antiques, and so forth. Sharp, pointed objects and objects that carry destructive or decaying energy will bring negativity into the area in which they are placed. I have seen homes with noose-like ropes as decorations in a family room!
- Make sure you do not place images that depict destruction, aggression, and violence in the home. These include paintings and photographs. A depiction of a massacre or a bloody battle scene will bring destructive energy into a space regardless of whether it is recognized as esteemed art or not.
- Indoor plants should not have pointed, gnarly, and spiky features. In general, leafy and flowering plants bring positive energy while plants like cacti and certain succulents can bring unfriendly energy. If you want roses, make sure the thorns are removed. Many florists now remove thorns to

make roses child friendly. Some may even carry "thornless" roses.

- Images of water, the color blue, can be used to counter the power of fire in the kitchen and in rooms with fireplaces. Place pictures of waterfalls or a bowl of water near the fireplace. Make sure the waterfall does not face a door to the outside or a large window. If this is not viable, fill a glass bowl with water and place it on the mantlepiece or on the floor close to the fireplace. A glass bowl will display the presence of water, and glass is an object that carries the energy of water. Kitchens have sinks and a constant presence of water. As long as your appliances and countertops are not bright red in color, there is no need to place any images of water in the kitchen.
- Images of fire should not be placed above the fireplace or above the headboard of a bed. Fire introduces volatility to the bedroom and a fire hazard to a room with a fireplace.
- Images that have red as a dominant color should also not be placed above the fireplace or in the bedroom. Examples are photographs that show red rocks or "flaming" autumn leaves.
- For the same reason stated above, red tablecloths, upholstery, bed covers and sheets, cushions, carpets, or lampshades should not be placed in the vicinity of fireplaces or on or near beds.

There is a reason why decorating your space is the last step in working on the fengshui of the interior of a home. Do not let ideas of decorations distract or override the more important fengshui factors such as neighborhood, surrounding land features, architecture, and floor plan. Don't let beautiful floor tiles in the kitchen make you ignore the bad fengshui of a road running directly into the cooking area.

The Income-Generating Unit

Income-generating units on a property help defray maintenance costs, pay off mortgages, add equity, and build savings toward retirement. From a financial point of view, an income-generating business such as a bed-and-breakfast is a profitable enterprise. However, when designing an income-generating unit that is attached to a home, we need to be careful to separate energy brought in by the guests from the domestic energy of the household.

Guidelines on Designing an Income-Generating Unit

- Ideally, the income-generating unit should be a separate structure and not attached to the home. Placing it above the garage is viable, if it is housing short-term guests. It is not recommended to put long-term rentals in rooms above a garage for the same reason you don't want to put your living space above the garage.

- If it is not possible to locate the income-generating unit in a separate space outside the home, then the unit should have a separate entrance.
- Basement rental areas are viable if the rental is short-term. For the same reason that it is not desirable for you to "live" in the basement, it is not desirable for long-term renters to live in basements.
- The income-generating unit should be smaller than the residential space of the owner. If the guest suite is too large, the guest will dominate or intimidate the owner.
- If there is common area used by the owner and renters such as the laundry facilities, this area should not be near the bedrooms, family gathering space, or the home office of the owners. Ideally, the guest should not need to walk through the home to get to the shared facilities.
- If the guest is to have breakfast in your kitchen, it is best to limit the service to a short period.
- Ideally, long-term renters should have their own kitchenette in their space and not use the owner's kitchen.
- Carefully plan the use of outdoor space. It is best if the renters do not use the same outdoor space and facilities as the owners. These spaces include patio and outdoor kitchens, pools, hot tubs, and so on. Shared use of swimming pools is tricky because of safety and liability concerns.

- Having housemates is a special form of sharing space. Whether you are renting rooms to other occupants or you are a renter of a room with other occupants, you need be careful not to let the energy of other occupants intrude into your private space. You may share kitchen, bathroom, and living room space, but don't share your bedroom space.

Fengshui Garden

Earlier you learned that some garden features can bring negative energy and that you should remove them or not install them at all. Mitigating harmful energy is concerned with protecting your home. It is about what *not* to do.

There is a form of garden fengshui that is designed to enhance the home, not just to protect it. Putting in a garden that enhances your home is not something you *have* to do. Doing it will increase positive energy around the home. However, not doing it will not harm you.

Garden fengshui is about installing improvements after fengshui problems have been removed. Do not work on enhancement features until you have mitigated or removed the problematic ones.

We all know that having a garden with colorful flowers and leafy beautiful trees will enhance a space. What I will show you below is an example of how you can use the five elements to bring in harmonious energy. The five elements arranged in a harmonious sequence are metal, water, wood, fire, and earth. In Chinese cosmology, this sequence is called the Creation Cycle, because each element creates and nourishes the element following it. Metal creates water because traditionally, water sources are associated with the presence of metal. This is the rationale behind using the method of dowsing to search for water sources. Water nourishes wood. This is obvious because plants need water for growth. Wood creates fire because wood is fuel for fire. Fire creates earth because ashes are produced after a burn. And earth creates metal because metal is born in the bowels of the earth. The sequence is cyclical, so the last element will carry the energy of nourishing to the first element.

There are two ways you can arrange this nourishing sequence of the elements in your garden. The first arrangement is circular. This sequence is best suited for large spaces like backyards. The second arrangement is linear and is best suited for narrow spaces such as side gardens.

Here are examples of garden features or ornaments that can be used to carry the energy of the five elements.

Metal: Metal garden ornaments like sundials, bird feeders, or metal chairs

Wood: Woody plants, wood chips arranged elegantly, wooden lawn furniture

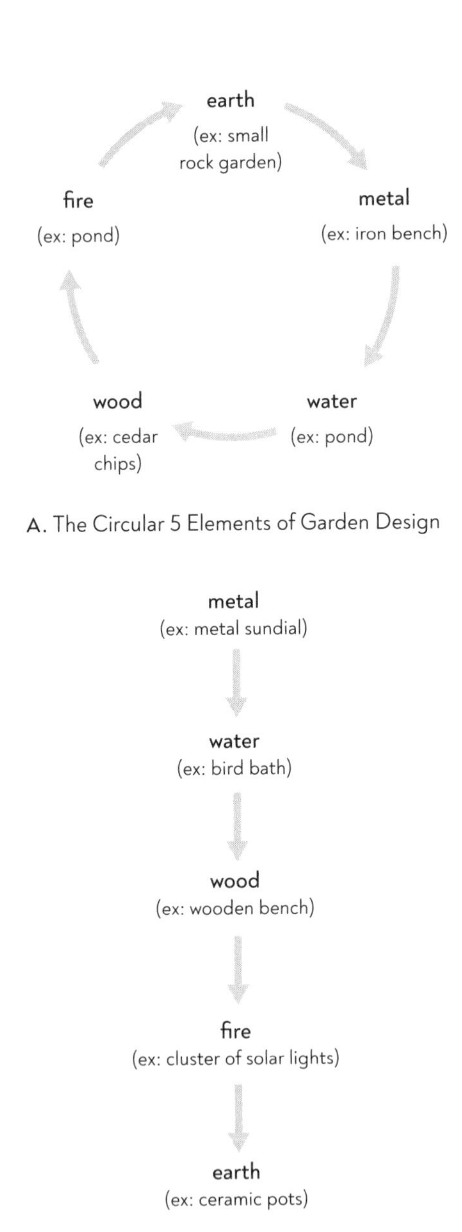

FIGURE 10. Garden fengshui: two kinds of five-element garden design.

Water: Ponds, small fountains, bird baths
Fire: Braziers, outdoor firepits, a bank of lights
Earth: Ceramic pots, rock features

Garden fengshui is fun to do, and it can bring elegance and nourishing energy into your side or backyard. You can get materials inexpensively, and it is a great project to involve the whole family.

PART THREE

Fengshui Strategies

Ideal fengshui of any space is rare. Most of the time we need to live with what we have or what is available. This is especially true when you own a property or are locked into a lease. Therefore, the use of countermeasures in fengshui is crucial to ensure that your home is safe and livable.

Countermeasures are designs used to deal with adverse conditions. They mitigate negative fengshui effects by helping us avoid, dissolve, remove, or weaken destructive energy. Countermeasures can be used to work with both external and internal environments, such as nearby land features, the architecture of surrounding buildings, the exterior architecture of your building, the floor plan, interior architectural features, and interior decor.

This section will give you some guidelines on how to design and place countermeasures in order to make a space that is safe and comfortable for you to live in. You can use these guidelines to work on a space you are living in or plan to remove small problematic features in a space you are going to move into.

13

General Fengshui Strategy

BEFORE WE BEGIN to design countermeasures, here are some general strategies you can use in dealing with adverse fengshui conditions.

First, before you reject a potential space or panic about where you are currently living, you should make a list of those areas you consider to have adverse fengshui, rate the severity of the negativity, and estimate the effort and cost needed to install the countermeasures that are necessary to neutralize them.

Then, take a look at the list below of the most severe negative fengshui conditions. Since most of them involve features in the external environment, it would be difficult or almost impossible to alter them.

A home has severe negative fengshui if it is located in one or more of the following situations:

- Where a road runs straight toward it
- Between thoroughfares, including roads, railroads, and bridges
- On a bridge-like platform over a river, gorge, or gully
- Perched precariously on a cliff side
- Directly under overhanging rocks
- Surrounded by knife-edge rocks

The strategy in these situations is to figure out how to move out of the home. It is very rare that I recommend that people leave a building. In most buildings, problems can be mitigated, some with more cost and effort and others with less. However, in the above conditions, the best strategy is to find another place to live.

Finally, always consider protection first. The health and safety of occupants are the most important factors in choosing a place to live. Wealth and enrichment are only possible if you are healthy and safe.

14

Using Countermeasures against Harmful Structures in External Environments

HARMFUL STRUCTURES in external environments include land formations, the architecture of surrounding buildings, and objects. In fengshui, these harmful features can be countered by strategic placement of items that will reflect, absorb, deflect, bounce, block, or destroy the harmful effect of these structures. In this chapter we will explore different types of countermeasures.

Reflector-Type Countermeasures

A reflector is designed to send the harmful effect back to the structure causing harm, thereby preventing it from entering a home. Mirrors, a large sheet of foil, or a basin or pool of water can all be used as reflectors. Reflectors are best used against precarious, falling, and colliding objects such as loose rocks, shiny objects like satellite dishes and solar panels, shiny surfaces of cars, and reflections off windows or pools. The incoming energy is reflected back at the object that is generating the reflection. Bouncers and absorbers will not work against reflecting objects.

When installing a reflector-type countermeasure, make sure that the structure you are trying to ward off is imaged in the reflector. While you can reflect small objects with a small mirror, you may need a large mirror or shiny aluminum sheet to reflect a large structure.

Absorber-Type Countermeasures

An absorber is designed to absorb or "eat up" the harmful effect of a structure and prevent it from spreading. It is like a pincushion absorbing needles that are pushed into it or a bed of cotton wool absorbing a solid object thrown at it. A sandpit or a bed of cedar chips can work as an absorber. A pile of gravel does not work because it is not "soft" enough to absorb the energetic impact of an incoming force.

Absorbers are best used against knife-edged rock slabs, spear-like rocks, large overhanging rocks, architecture or sculptures that have sharp and pointed shapes or those that resemble spears or arrows, pergolas with horizontal beams resembling spears or battering rams, trees with branches like spears, and small sculptures resembling weaponry. For the absorber to be effective, it must be angled at the harmful structure. The size of the absorber is dependent on the size of the harmful structure aimed at the home. The absorber countermeasure is best used against harmful structures that are positioned higher than the home.

Deflector-Type Countermeasures

A deflector is designed to redirect the path of harmful energy away from a building. While a reflector will reflect the incoming energy back to its original location, a deflector, when aimed strategically, can redirect the harm to another location. A system of reflective surfaces can be used as deflectors if the angles are adjusted so that the incoming force is redirected to a specific location. Deflectors are tricky to install because if you err in angling the deflector, the negative energy can be sent to a neighboring home and harm the occupants. This is why the placement of deflectors requires precision. The incoming harm must be deflected accurately back to the object, toward an empty lot, or toward a large body of water. You also want to make sure it is only the energy of the harmful object that is deflected, not any adjacent positive energy. Don't deflect the elegant flower bed along with the garbage pile. If you are uncertain about how to aim the deflector, it is best to refrain from using this device.

Deflectors are best used against small sharp objects. To deflect a large object, the size of the deflector must

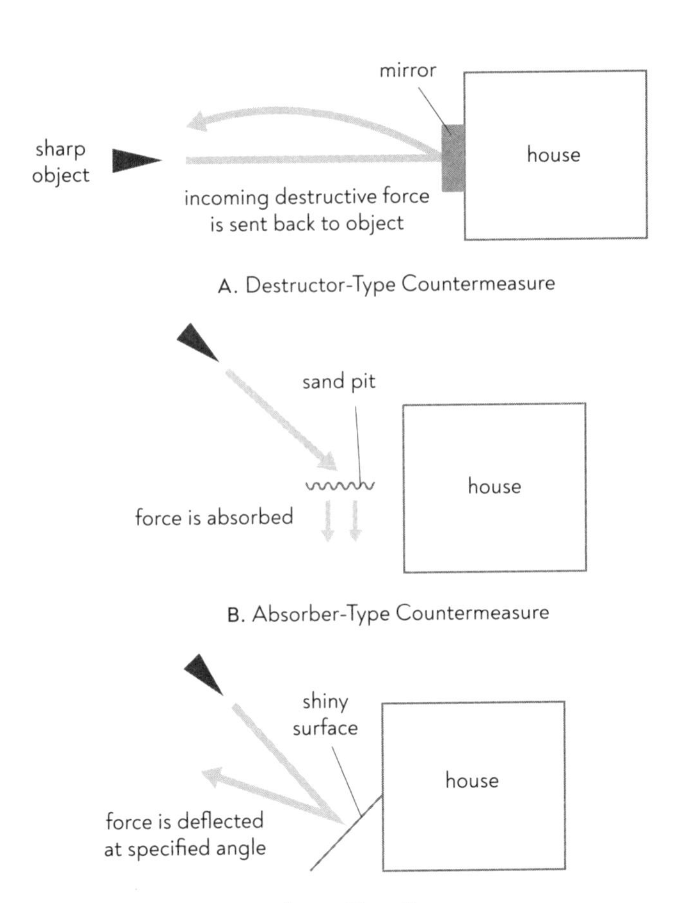

FIGURE 11A. The use of destructor-, absorber-, and deflector-type countermeasures.

be at least as large as or larger than the harmful object. Since it is not realistic to install a large deflector, this countermeasure is only useful against small objects such as antennas, small garbage piles, or dumpsters.

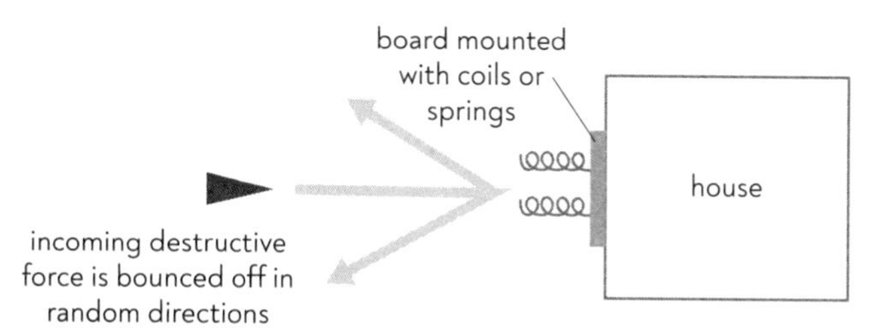

A. Bouncer-Type Countermeasure

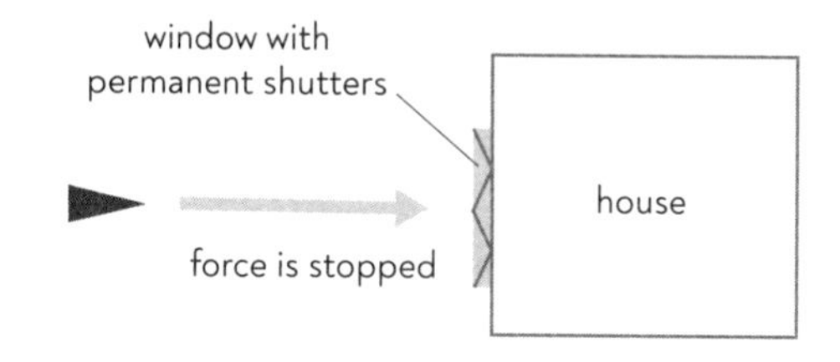

B. Blocker-Type Countermeasure

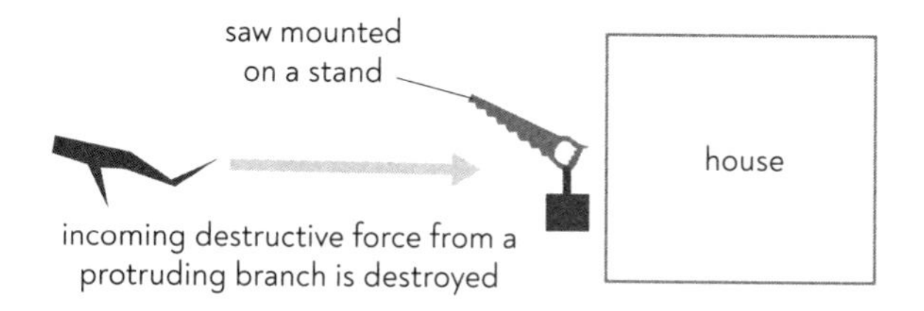

C. Destructor-Type Countermeasure

FIGURE 11B. The use of bouncer-, blocker-, and destructor-type countermeasures.

Bouncer-Type Countermeasures

A bouncer is designed to bounce the effect of a harmful structure away from a building. There is no way of controlling where the incoming harmful energy will "bounce" toward: it is simply bounced off somewhere else. This countermeasure should not be used if there are neighboring buildings or public spaces such as parks because the harmful energy can be bounced unpredictably to those spaces and harm the occupants. Effective bouncers include coils of springs, a baseball bat, or boxing gloves mounted on the ends of sticks. If it is not possible to install these objects physically, you can use images. However, images will be less effective than physical objects.

Blocker-Type Countermeasures

A blocker is designed to stop an incoming harmful effect dead in its tracks before it can enter a building. Blockers work like a closed door or brick wall. Solid doors, boarded-up windows or windows with permanent shutters, solid walls, or thick hedges will work. The key to this countermeasure is making sure that there are no cracks in the blocker. Otherwise, the incoming negative force will find a way to get through.

Blockers are best used against harmful energy rushing toward you, such as driveways and roads pointing at your home. However, blockers can also have undesirable effects. For instance, you can deprive your home of light by permanently closing the shutters.

Destructor-Type Countermeasures

Destructors are the most powerful of all countermeasures, designed to destroy the incoming harmful energy.

Typically, the countermeasure is an object of destruction itself. They can be real objects such as arrowheads, knives, saws, or hammers, or they can be miniature representations of destructive objects. Models of cannons, guns, armored vehicles, or rockets can also be used. If you are unable to acquire real or model objects of destruction, you can use images of the above items.

For destructors to work, they must be energetically stronger than the force they are trying to destroy. For example, if there is a sculpture resembling a gun in a park across the street, using a knife as a destructor will not work. In this situation you had better use a miniature cannon or a model rocket. Against electrical transformers or high-voltage wires that hang over your backyard, you can bury a pair of scissors underneath the wires with the scissor tips pointing up. For safety, the scissors can be mounted inside a box. Some destructive objects simply cannot be countered by any destructor you fabricate. If your home is located across from a war memorial park with a tank or artillery pointed at you, it's time to move.

In terms of effectiveness, the real object is most effective, followed by a model or miniature representation. Images are the least effective. If you aim your countermeasure toward the destructive object threatening your home, the countermeasure will not bring negative energy into your environment. This is akin to shooting an arrow from a bow. The bow will not harm you, and the arrow is shot toward the target. Make sure the destructive-type countermeasure is always "aimed" toward a target.

Arrows, models of guns, cannons, and so on, will work. Do not use an image of an explosive device like a firecracker. Not only will this device be unable to neutralize the threat, but it will actually carry destructive energy inside the home.

Summary of Countermeasure Usage

Reflectors and deflectors can be elegantly designed and installed but their placement must be precise. Absorbers are easy to install but are limited in use. Blockers are effective but are not always elegant or practical. Boarded-up or shuttered windows are not pretty to look at. They also block out light that is needed to let nourishing energy into a space. Bouncers are tricky to apply, and should be installed with consideration of neighboring buildings and public places such as parks and playgrounds. Destructors are the most powerful countermeasures. However, if they are used carelessly and frivolously, they can introduce aggressive energy into your home and the neighborhood.

Designing and applying countermeasures is an art. The more creative you are, the more ways you can figure out how to counter adverse structures in the environment around you.

15

How to Use Countermeasures

AS WE SAW in the last chapter, countermeasures are devices that are designed to neutralize threats to a building and its occupants. Not all spaces are perfect. There are usually some problems that need to be mitigated before we can live and work there safely and comfortably.

Some issues are associated with features in the external environment such as land formations, architectural features from neighboring buildings, and objects in the vicinity of a home. Other problems are associated with features in the internal environment. These include floor plans, interior decor, and indoor architectural features such as beams, pillars, and so on.

Certain problems can be easily mitigated by moving some furniture, painting a wall, or changing window coverings. Others require more time, effort, and resources, such as removing or adding walls or repositioning stairways. Then there are problems that can be mitigated by installing special devices to redirect the energy of destructive objects away from a home.

In this chapter, you will learn how to design, build, and use countermeasures to make your space safe and comfortable to live in.

However, there *are* some problems that cannot be mitigated by countermeasures. These include massive rock formations, adverse road patterns such as a T-junction, and large buildings that have overwhelmingly aggressive architecture. The only solution in these situations is to find another place to live, or, better yet, not to move in at all.

Against Adverse Road Patterns

Harmful road patterns are best countered by blockers and reflectors.

Countermeasures to Mitigate or Minimize the Negative Effects of Road Patterns

- Types of roads pointing at your building
    - **A neighbor's driveway:** Use a blocker. Hedges are best. Make sure that the hedge is sufficiently wide to block the view of the driveway from a window inside the home. Hedges can be elegant, and against a driveway, they should work.

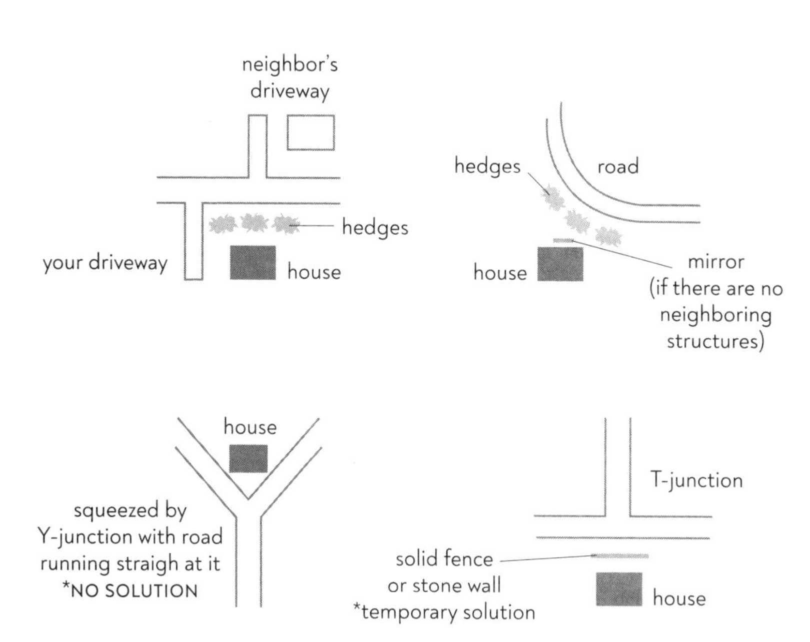

FIGURE 12. Placement of countermeasures against harmful road patterns.

- **A curving road:** Use a blocker. If there are no neighboring structures, you can use a reflector like a mirror to reflect the energy of incoming vehicles.
- **A residential street:** You should again use a blocker. Soft blockers like hedges may not be sufficient. A solid wooden fence is recommended.
- **A T-junction of a busy road:** A solid wall in front of the home is needed. However, this

countermeasure is only a temporary protection until you move out.

- A building squeezed between a Y-junction: This cannot be rescued.

Against Problematic Floor Plans

Problematic floor plans can be mitigated by removing walls, repositioning doors and windows, and installing materials that can reflect or absorb negative energy.

Some mitigations are only feasible if you own the property. It is unlikely that the owner of a rental property will modify floor plans for a tenant. However, there are some countermeasures that renters can use against problematic floor plans, although their effects are fewer compared to remodeling.

Problematic Floor Plans and How to Counter Them

- **Stairs that flow toward doors to the outside:** If you own the property, the ideal mitigation is to relocate the position of the door. Reconfiguring or changing the location of the stairs is more difficult architecturally and more expensive.

  If you are a renter, or if changing the location of the door is not feasible, you can place a waterfall fountain at the bottom of the stairs and aim the flow of water toward the stairs going up. The momentum of the waterfall energy will serve to counter energy rushing down from the upper floor toward the door, thus minimizing health and wealth energy exiting the home. Another option, less effective than the physical presence of water, is a picture of a waterfall with the direction of the water aimed up the stairs. The more realistic the image of water, the more effective will be its power. Photographs are best, followed by realistic paintings. Abstract images of water will not be effective.

- Long narrow corridors: If you own the property, the ideal mitigation is to widen the corridor by removing some walls fully or partially. For example, the wall can be partially removed and replaced by a glass plane. This gives the feel that the corridor is no longer enclosed because it is now visually connected to another space. This mitigation is not practical if a bedroom is behind the partial glass pane. However, it is viable if the room exposed is a home office.

  If you are a renter, or if opening walls is not feasible, then you can create the feel of spaciousness by lighting up the corridor. Either ceiling lights or wall lighting will work, but to maintain the effect of spaciousness, the corridor needs to be lit, especially when there are occupants inside the home. If the paint color of the corridor wall is dark, you can repaint it with a light and bright color. Anything to minimize the darkness of a narrow corridor will make it feel more spacious.

- **Small confined spaces:** Whether you own or rent the property, an easy way to decrease the claustrophobic energy in the space is to brighten it with light-colored paint and light-colored furniture. Dark-colored paint and dark wood furniture will make the space feel even smaller and more constricted. Another way to introduce space into a small room is to have smaller pieces of furniture. Large beds, desks, and cabinets will crowd the space, making it feel extra confined.
- **Floor-to-ceiling windows:** Whether you own or rent the property, an easy way to reduce the unwanted exposure to the outside and minimize energy escaping or negative energy entering is to place low furniture against the lower sections of the window. Furniture such as low bookcases and planters can block undesired exposure. Frosting the lower section of the windows will also work. You can mimic the effect of frosting with appliqué sheets that are easy and inexpensive to install and remove.
- **Garages below living space:** If you are happy with parking the cars on the street or in the driveway, you can turn the garage into a storage area or a hobby workshop regardless of whether you own or rent the property. If you own the property, you may want to consider converting the garage space into a room. This will completely mitigate the problem of negative energy entering a room from the garage below.

If garage conversion is not feasible, a little replanning of space usage can turn a bedroom or home office above the garage into a recreational space. Energy in rooms above garages is unstable, making them unsuitable spaces for bedrooms or home offices. However, the space is viable for recreational activity. If you make the room a guest room, make sure the guests do not stay longer than a month, or they could be impacted by the negative energy.

If alternate use of space above a garage is not feasible, there are countermeasures you can use to minimize the effects of instability caused by vehicles moving in and out of the garage. These mitigations will minimize the problems; they will not totally remove them. To keep the unstable energy in the garage from flowing into the room upstairs, you can get sheets of shiny material and fasten them on the ceiling of the garage. Foil, reflective film, and thin panels of mirrors can be attached on the garage ceiling. These materials are inexpensive and easy to install. Their reflective nature sends the energy in the garage back to where it was generated, keeping most of it away from the room above.

- **Garages directly adjacent to a living space:** To minimize energy from garages leaking into a living space, you can use the same method described above. Simply place the reflecting material on the wall of the garage shared with the living space behind it. This will send most of the energy back into the garage and keep it from leaking in.

One last bit of advice: Countermeasures are often used when you cannot relocate. The best way to deal with problems is not to have problems in the first place. A good strategy to consider when you are looking for a space is to make sure that you don't have to install countermeasures in the first place. Don't choose a home based on kitchen and bathroom fixtures and then find out that you have to install massive countermeasures. See if you need to install countermeasures and determine the scope of the mitigations before you choose a home.

Against Adverse Interior Architectural Features

Interior architectural features can have a large impact on the fengshui of a home. Some can be mitigated easily but some need time, effort, and ingenuity.

Mitigating Harmful Interior Architectural Features

- **Harmful light fixtures:** Harmful light fixtures are those with sharp features pointing down or spreading outward. These include metal and wooden fixtures resembling spikes and chandeliers resembling shards of broken glass. These fixtures will "attack" health if they are above a bed, threaten livelihood if they are above a desk or a dining or cooking area, and cause conflict among the occupants if they are above sitting areas in the family room. They may also increase the likelihood of occupants being physically injured, whether inside or outside the home.

The best way to mitigate harmful light fixtures is of course to remove and replace them with benign-looking designs like domes, round globes, or even canned fixtures. Even if you are a renter, it is possible to negotiate with the owner to change a light fixture.

If it is not possible to change a light fixture, you can cover them with a large dome-like shade. A large paper lantern shade or a round mesh cage would work. You can even fabricate these yourself. You can soften a rack of metal or wood spikes by weaving some artificial vine-like plants around the rack of spikes, thus camouflaging it so that the spikes are no longer prominent.

- Harmful ceiling fans: Fans have blades, whether they are turned on or off. Blades above a bed are associated with illness and injuries. Above a cooking and eating area, or desk, blades are associated with forces that "cut" into livelihood. Fan blades above a social gathering area are associated with conflict and quarrels.

  The best way to mitigate harmful ceiling fans is to remove and replace them with fans that have harmless features. You can now find ceiling fans that have blades hidden inside enclosures like boxes, mesh, or cages.

If replacement of ceiling fans is not possible, you can replace the harmful-looking blades with leaf- or frond-like blades. If this is not possible, you can fabricate cloth cover with a leaf pattern to cover the blades, thus making them look like fronds instead of blades.

The only time when the negative effect of a ceiling fan is minimal is when it is installed on a high ceiling, usually a ceiling that is at least twice the typical ceiling height.

- Sharp, rough, and rocky facades of interior walls: Sharp and rough rocky interior walls carry aggressive and heavy energy. They also tend to make the space claustrophobic. The best way to mitigate these harmful facades is to remove and replace them with a flat wall, tile, or slate that is not red in color. If these options are not possible, then paint the rocks white or another light color. A light color will diminish the aggressive and heavy energy carried by dark rough rock.

  If the rough heavy-looking rock facade is above a fireplace, then it is not advisable to cover this surface with hanging plants. Plants carry wood energy, and if they are near the presence of fire, they will exacerbate the harmful effect of fire even if the fire is a covered gas or electric fireplace rather than an open flame. However, on walls not adjacent to a fireplace, the rocky facade can be covered with real or artificial vine-like plants, making the wall appear like a "living" wall covered with greenery.

If you are a renter and the owner is not amenable to you painting the rocky facade above a fireplace, it is better to choose another place to live.

- **Ceiling beams:** Beams carry the energy of oppression. If there are beams over the sleeping, dining, cooking, or work areas, the negative energy of the beams can be mitigated by covering them with cloth. Simply wrapping an elegant print around the beam will work. If a beam is above a bed, an option is to construct a canopy above the bed so that the beam is "covered" by the cloth. Install two curtain rods on the ceiling and fasten a cloth onto both rods to make a canopy.

  Painting the beam the same color as the rest of the ceiling will minimize its negative effect, but it is not as optimal as covering it.
- **Oversized fireplaces:** Oversized fireplaces introduce an unusually strong energy of fire into a room. Since fireplaces are difficult to remove or relocate even if you own the home, the best way to "decrease" the presence of the fireplace is by painting the mantle and the facade the same color as the surrounding wall. If the fireplace is a large wood-burning structure, it is also possible to

change it to a smaller electric or gas insert. Install a smaller-sized insert and fill the empty space with fire-resistant material. In this way, you have basically "shrunk" the size of the large fireplace.

Against Adverse Interior Decor

Problematic interior decor is the easiest to mitigate. It does not involve structural changes or large-scale remodeling.

Mitigating Problematic Interior Decor

- **Paint colors:** Red-colored paint carries the element of fire, even if it is only an accent wall. Because fire is a strong element, a small amount of red will have a larger effect than a small amount of blue or green. A red wall in a bedroom is associated with fiery relationships. In a kitchen it is associated with a fire hazard. In a dining room, it is associated with conflict during dinner. In a home office, it is associated with business quarrels. In a family room, it is associated with conflict among family members.

  Light wall color is preferable to dark colors. Dark colors tend to introduce heavy and lethargic energy. Dark-colored paint also absorbs light, making the room dark and gloomy.

  The simplest way to mitigate excessive amounts of red and dark-colored paint on the walls is to repaint with colors like green, blue, earth tone tints, or even white. Even if you are a renter, owners are usually amenable to tenants changing paint color, especially to colors that are more neutral.

- Wall coverings: Wallpaper carries energy similar to paint colors, but the pattern of the wallpaper can introduce additional elemental complexities. The effect of wallpaper with a solid color is essentially the same as paint color. However, patterned wallpapers can carry energy that is specific to the patterns. For example, a wallpaper with a forest scene or flowers will introduce plant energy. A wallpaper with geometric patterns will introduce energy associated with structure and organization. Busy patterns introduce busy energy. Relaxing patterns introduce relaxing energy. Patterns with sharp and pointed elements introduce conflict and aggression. Dark patterns introduce dark and lethargic energy.

  The safest wallpaper patterns, if you wish to apply wallpaper or retain what you have, are ones with simple, relaxing images in a light color. Wallpaper is easy to remove if it turns out to bring negative energy. If you are renting a property, most owners are amenable to tenants removing and/or replacing wallpaper.
- Window coverings: Window coverings often occupy much surface space, especially if the win-

dows are large. Curtains and shades can introduce either positive or negative energy into a space.

Dark-colored and busy-patterned curtains will introduce the same kind of effect as dark-colored paint or wallpaper. The easiest way to mitigate these problems is to install new curtains that are light and neutral in color.

Window shades not only carry color-related energy but also introduce additional complexities. You can replace dark-colored shades with light-colored ones. However, some shades are shaped like blades. Window or glass door coverings that have vertical shades will introduce blade-like energy into the space. If you cannot remove the vertical shades, simply keep the shades permanently moved to the side, install a curtain rod above the window, and hang curtains. A similar solution for blinds that cut horizontally is to keep them up and install curtains over them.

- Flooring: A floor is a large surface area, and large surface areas can have a big impact. Most flooring tends to be neutral in color. However, some floor coverings, especially tiles, may be bright red or have a pattern that is too busy. Terra-cotta tiles tend to be red, and if they cover an entire area of a space like a kitchen or an area near a fireplace, it can lead to a fire hazard.

  A busy tile flooring introduces busy energy, even if it is in the bathroom. If an area has "busy" energy, you will be preoccupied with a lot of activities, especially unnecessary ones. Life is complicated enough. You don't need added unnecessary complexities.

If you cannot replace an existing flooring, the next best option is to cover areas of the floor with a mat or carpet so that only a small area of the original flooring is visible.

- Decor objects: Decor objects are either functional or are there to make a space more elegant. Objects such as art, statues, ornaments, books, standing lamps, and so on are often displayed to make a space more elegant. If an object carries harmful energy, remove it. The general rule of thumb about decorative objects is to make sure that they do not carry images of or represent objects of destruction. Art and statues that depict war, massacre, or violent action should not be in a home. Large objects that are predominantly red in color, such as seat covers, cushions, and lamp shades introduce strong fire elements. The simplest way to mitigate problematic decor is to remove or cover it.

  Functional objects are not considered decorative objects, but because they are functional, some of them cannot be completely removed.

  First and foremost are mirrors. Mirrors are functional. However, mirrors reflect energy, and if they face a window or a door to the outside, they will reflect both health and livelihood energy to the outside. Relocate the mirrors to a wall that does not face a window or door. Mirrors facing beds will take energy away from those sleeping there. The life energy in the room will be divided to serve two instead of just one of you. It should be easy to relocate the bedroom mirror to a wall that does not face the bed.

Secondly, if you have large kitchen appliances and cabinets that are red in color, you will need to mitigate this problem immediately. Too much red in the kitchen will increase the likelihood of a fire hazard and burn accidents.

If you have red-colored cabinets or a red refrigerator, you will need to paint the cabinets a different color or apply an applique to the refrigerator. Small red-colored appliances such as toasters, mixers, juicers, and so forth, can be put inside a cabinet or pantry and taken out only when in use. This reduces the presence of red objects on the countertops.

Problems with interior decor are the easiest to mitigate. Therefore, they should not be the criteria for you to reject a space. If everything else about the home, exterior and interior, is not problematic, you should not let interior decorative objects prevent you from choosing the space. You can deal with decor objects after you move in. If you are already settled into the home, use the guidelines above to mitigate problems associated with interior decor and design.

16

Folk Fengshui Remedies

THERE ARE FORMS of fengshui "remedies" that are based on folk beliefs. They are not considered fengshui "countermeasures" in the strict sense but are often used to bless and protect a home.

Chinese Folk Fengshui Remedies

Chinese folk fengshui remedies include supplicating spirits and deities to bless and protect a home, securing the good graces of the ruling deity of the year, and installing good luck charms, objects of power, and talismans in and outside a home.

Spirits and Deities

THE TAISUI

The Taisui is the guardian deity of a lunar year. It is

actually the planet Jupiter. Sometimes called the Grand Duke, the Taisui is a system of sixty guardian deities, each guardian ruling a year within the sixty-year Great Cycle of the Chinese calendar.

The Taisui is believed to be a powerful force that affects luck. If you are in its good graces, you will be blessed. If you offend it, you will have bad luck. Each year, the Taisui occupies a certain position in the eight directions of the compass. It is said that if you do construction or dig up earth in the direction the Taisui is occupying, you will "offend" the deity, which could bring bad luck and even retribution from the Taisui. For example, in the year 2022, the Taisui's position is in the northeast. The belief is that if you renovate a space or dig up earth in the northeast, you will disturb the Taisui, which will bring bad luck.

In Chinese culture, this belief has affected people's decisions to avoid making renovations in the direction of the home the Taisui is occupying. However, on the practical side, it may not be viable to stop major construction projects altogether in a direction occupied by the Taisui. In my construction business, we make an offering to the Taisui of the year, apologize for the inconvenience, and go ahead with the work.

How do you find the Taisui's position for a certain year? Thanks to the plethora of information available through the internet, you can get this information by searching for the "Taisui of the year."

KITCHEN GOD

The Kitchen God rules all activities in the kitchen. He blesses the hearth, makes sure a family has enough to eat, and protects the occupants of a home from illness associated with food.

The Kitchen God is also a watchdog of the activities of all household members. He documents their ethical and unethical deeds and reports to the higher deities at the end of each year.

Installing a shrine to the Kitchen God in the kitchen is a way of asking him to bless and protect all matters of livelihood and health associated with food. You can purchase a Kitchen God shrine in Chinese markets and online Chinese gift stores. The shrine should be located in the kitchen on a shelf.

If it makes you feel blessed and protected, there is no harm in installing a Kitchen God shrine. However, don't rely on the Kitchen God alone to protect you from a road running straight at your kitchen.

Make sure the kitchen is in a protected area of the home and mitigate any harmful effects using the countermeasure strategies described in this book before you consider asking for extra protection and blessing from the Kitchen God.

DOOR GUARDS

Door Guards can help prevent destructive energies from entering a home. Door Guards come in pairs and must be placed outside the home in order for them to be effective protectors. You can place images of Door Guards on the doors of the main entrance or on the wall on both sides of the entrance. Statues of Door Guards should be placed on pedestals outside the main entrance. You can purchase images and statues of Door Guards in Chinese markets and online Chinese gift stores.

Don't rely on Door Guards alone to protect you from an aggressive architectural feature across the street or a neighbor's driveway pointing toward your home. Install the relevant fengshui countermeasures first. Then you can think about getting extra protection from the Door Guards.

EARTH GOD (TU DI)

The Earth God, or Tu Di, is the guardian of the grounds of the home. He blesses and protects the home and the occupants that live on the property. The most important function of the Tu Di is to make sure that the occupants are not harmed by accidents or mishaps when they are on the property.

The Earth God shrine is always placed on the ground. The shrine can be placed just inside the main entrance or against a wall on the kitchen floor. If a home has a garden, the Tu Di shrine can be installed there.

If it makes you feel more safe and protected, there is certainly no harm in having a Tu Di shrine on your property. However, don't rely on the Earth God alone to protect your home from aggressive-looking features across the street or volatile energy from electrical transformers near your home. Make sure you have installed the relevant fengshui countermeasures first before you consider installing an Earth God shrine for extra protection.

Magical Objects and Luck Charms

Good luck charms and objects that carry magical qualities are often used to bless and protect a space.

THE BAGUA

Most popular magical object is the Bagua, a pattern of eight trigrams associated with the eight directions of the compass. The Bagua is said to have a magical quality capable of warding off destructive energies. A plaque with a Bagua pattern is typically placed facing a direction where destructive forces are expected to enter.

A variation of the Bagua is a Bagua pattern with a mirror in the center. The mirror is said to give additional power in reflecting incoming destructive energy away from the home. The Bagua mirror can be considered a reflector-type of countermeasure with the added folk magical power of the trigrams.

Bagua plaques and Bagua mirrors can be purchased inexpensively in Chinese markets and online Chinese gift stores.

STRING OF COINS

Another good luck charm is a string of old Chinese coins. These are hung especially in the home office to enhance wealth. However, for this charm to work, the office must have a distant view that inspires you to create and maintain wealth. In offices without views of a horizon, it is unlikely that a string of coins will enhance your wealth.

To make a string of coins, get some old Chinese coins, tie them together with a red cord, and hang the string on the wall of your home office.

TALISMANS

Talismans are writs of power. They can be etched on a metal plate or written in red ink on yellow paper. In Taoist magic, there are talismans that can protect you from harmful forces, enhance your health and wealth, and bring harmony into a home.

Talismans can be hung anywhere in a home. Usually talismans of protection are hung near doorways and windows; talismans of health are hung in bedrooms and kitchens; talismans of harmony are hung in bedrooms; and talismans of wealth are hung in home offices.

You can obtain metal wall-hanging talismans from Chinese gift stores. Paper talismans are written by Taoist priests and can be obtained in Taoist temples.

Talismans are not considered fengshui countermeasures in the strict sense but can be used to give added blessings and protection *after* the home is deemed to be free of fengshui problems.

Folk Blessings and Protections from Other Cultures

When using folk fengshui remedies, it is not imperative that you use only those from Chinese culture. All cultures have good luck charms, magical objects of power, and spirits and deities that will bless and protect you. If you are looking to augment your fengshui countermeasures or wish to get extra blessings and protection, you can use charms and blessings from a culture that you feel you are a part of. The most important factor in choosing a folk blessing or remedy is that you believe them and can relate to them meaningfully. Don't use or install a folk remedy from a culture you cannot relate to.

Advice on Using Folk Fengshui Remedies

Strictly speaking, folk fengshui remedies are not considered fengshui "countermeasures" and are not specifically designed to counter adverse fengshui. They could, however, bring additional blessings, luck, and protection after you have followed all the fengshui advice described in this book. You should never feel obligated to install shrines and magical objects for blessings, luck, and protection. However, now and then we can all benefit from an extra bit of luck in our lives. Just in case the hedge blocking an aggressive architectural feature from a neighboring building is not 100 percent effective, your Bagua mirror or the Door Guards are there to lend a helping hand.

17

Strategies for Choosing a Fengshui Consultant

WHEN WOULD YOU WANT to hire a fengshui consultant?

My suggestion is that if you have tried the recommendations described in this book, and you still have insurmountable problems, you may wish to hire a fengshui consultant. Choosing a fengshui consultant is like choosing a contractor to build or renovate your home. It is best if you can interview them before you use their services.

Tips for Choosing a Fengshui Consultant

- The best way to choose a fengshui consultant is to get referrals from people you know who have gotten a satisfactory consultation. Most of my clients come through referrals, whether they are for residential, commercial, or institutional fengshui. Some approach me after reading my books.

- Fengshui is not just interior design. If a consultant informs you that they will help you coordinate colors and work on interior designs, they are not fengshui practitioners. You can work on decor yourself.
- Some fengshui consultants will tell you to move walls or recommend massive renovations. Most fengshui problems can be solved without complex renovations. In fact, almost 99 percent of fengshui issues can be mitigated without opening a roof or adding or removing large sections of a house.
- A good fengshui practitioner will spend time discussing the fengshui issues with a client without committing to working on a space. Have an interview with a prospective consultant the same way you would with a building contractor. My fengshui colleagues and I always ask a prospective client to send photographs of the home, surrounding environment, and even floor plans before we give an estimate and the scope of the work we plan to do.
- Never hire a fengshui consultant to do a job without a written cost and time estimate.
- There are many fengshui schools. I recommend you hire someone who is trained in the traditional way. You can find a brief description of the traditional Chinese fengshui schools highlighted in

chapter 18. The techniques used by these schools have been tried and true for generations. Your fengshui consultant should use a method that has shown reliability for centuries rather than something that was invented ten or twenty years ago. The rationale for using an approach that's been around is this: If something does not work, it will not be around for long. If a system has been around for hundreds of years, then there is a good chance that it works.

- If you hire a consultant who claims to practice the Flying Stars technique, ask them to prepare a Flying Stars chart and walk you through every room in the home and evaluate its fengshui. Flying Stars is a complicated method, and it is used primarily to obtain additional information on the energetic flow of a space after you have determined that there are no substantial external, internal, or architectural issues. In other words, it is the next level of fengshui evaluation. You can do Flying Stars system yourself. My book *A Master Course in Feng-Shui* will show you how to learn and apply that method.
- There are some fengshui consultants who will tell you to move out if the direction of the home does not align with your birth year. I have one recommendation: *don't* take this advice. Fengshui is not that trivial. In all my years of experience as a practitioner of both residential and commercial

fengshui, I rarely recommend people move out of a place. And if I do, it is not because their birth year and building direction don't agree. It is usually because the external environment cannot be changed.

- The vast majority of fengshui problems can be solved to a certain extent. Ask your prospective consultant how they would solve fengshui problems that you have already solved. By the time you've gone through this book, you should be quite savvy in dealing with typical fengshui issues.
- Large-scale commercial and institutional projects benefit most from having a fengshui consultant. I have worked with banks, corporate headquarters, religious institutions, universities, and even government buildings. By all means hire a professional consultant if you are involved with these large-scale projects.
- Be careful of advice from a fengshui consultant who casually tells you to remove walls. Fengshui consultants are not necessarily contractors or builders. Making structural changes without advice from professional builders is dangerous. If a fengshui consultant recommends removing walls and other structural modifications, don't do it until you have consulted with a licensed contractor and/or builder.

I own a construction and engineering business. This allows me to make recommendations regarding structural changes with advice on whether engineering issues need to be addressed. I will never make a fengshui recommendation if the structural changes are dangerous or impossible to implement.

- My final advice is this: don't hire a professional consultant unless you have a lot of problems that you can't solve after you have read this book and after you have learned the method of Flying Stars from my *Master Course in Feng-Shui*. Trust your judgment and ability. Don't hire someone unless you are absolutely lost. This may sound strange since I am a professional fengshui consultant. I believe that professional consultants like me are most useful in large-scale projects. I would rather people do their own fengshui and not hire a consultant unless it is absolutely necessary. Chances are that you can do a better job yourself than the person you hire casually. Save money; don't hire us frivolously.

18

Fengshui Schools

YOU MAY WANT to learn a bit about different traditional fengshui schools before you hire a consultant.

The oldest school of fengshui is known as the Land Form school (in Chinese, Kanyu). It is believed to have originated in the third century. Focusing on evaluating how land carries energy, it is used primarily in fengshui projects that involve erecting large numbers of buildings on a piece of land. In China, Land Form fengshui is used extensively in the design and building of monasteries, systems of temples, palaces, and large mansions. I have used this system of fengshui to consult on the restoration of monasteries and temples in Hong Kong, Taiwan, and China. The major principles and applications of this school can be found in my book *Feng-Shui: The Ancient Wisdom of Harmonious Living for Modern Times*.

The next oldest school is the Three Periods school (in Chinese, Sanyuan), which originated in the seventh century. This school evaluates land features in conjunction with a complex geomantic compass consisting of several rings. This school is for nerds and serious practitioners who can read classical Chinese. The classics of Three Periods fengshui have yet to be translated. Instruction in this school is offered in fengshui classes usually based in Hong Kong and Taiwan. I have used this method to site and design large homes for people who have an idea of what this school can offer. This is an extremely complex method, and nobody with a practical sense would want to spend years learning it since it is used rarely.

The Three Combinations school (in Chinese, Sanhe) was founded sometime between the tenth and eleventh centuries. It is also a compass-based school that uses special rings in the geomantic compass. This school is just as complex to learn and apply as Three Periods school. It is not well-known, even to fengshui nerds and the traditional Chinese practitioners. The texts have yet to be translated. The Three Combinations method is best used in areas with dominant water features such as lakes and rivers and is only applicable if you are building a large complex in an area dominated by water. I only use it to situate and design a town that has views of water. It is not very useful in urban environments, and chances are that you won't need it to evaluate the fengshui of any urban residential and commercial building.

The Flying Stars school (in Chinese, Xuan Kong) was systematized in the sixteenth century but was said to have originated in the eleventh century. This is the most popular and most useful method. It was designed when China underwent massive urbanization, and more people with means began to live in cities. The Flying Stars school uses a basic compass system based on twenty-four directions. Using direction and temporal energy (the year when a building was erected), a chart called the Flying Stars chart is calculated and superimposed onto a space. The numbers of the chart describe where positive, negative, and neutral energies are located. I use this technique extensively when working with urban spaces. The method is used by many professional fengshui consultants in Asia, the Americas, and Europe. Flying Stars fengshui is not complicated, and with good instruction, you can learn it sufficiently to work on your home and even your business office.

There is a modern school of fengshui called Eight Mansions (in Chinese, Bazhai). This school was founded in the eighteenth century. It is easy to learn, but its use is restricted. It gained popularity for a while and then was abandoned by the serious professional consultants. This school proposes that the facing direction of a home needs to be aligned with the birth year of the head of household. If the match is not auspicious, then the occupants need to find a home that is aligned with their birth year. There are a couple of problems with this school. First, fengshui is not so trivial that a mismatch of the birth year of the head of household and the direction the home faces would warrant moving out of the space. Second, what if there are several heads of household? In many families, both spouses or life partners are equal breadwinners. Whose birth year should be aligned with the facing direction of the home? Then there are homes shared by multiple housemates. In this situation, whose birth year should be matched with the home's direction?

In my fengshui consultation, I use all the above techniques except for Eight Mansions. I believe in using the most effective technique that gives my client the most practical and least expensive way of solving their fengshui problems. Before you hire a consultant, you should first use the commonsense fengshui described in this book. The fengshui of a space is most affected by surrounding land and architectural features, followed by the architecture, floor plan, and interior decor of a space. If there are no problems in regard to these factors, then you may wish to use the method of Flying Stars to get more detail on how the energy is laid out specifically in each area of the home. If there are problems in the land, architecture, floor plan, and decor, no amount of Flying Stars knowledge will be able to mitigate them.

If you want more information about these traditional schools of fengshui, you can go to my two other fengshui books: *Feng-Shui: The Ancient Wisdom of Harmonious Living for Modern Times* and *A Master Course in Feng-Shui*.

19

And Finally . . .

WHEN I FIRST LEARNED traditional fengshui at the age of nine, I took an oath in front of my teachers, who happened to be my great uncle and his friend and colleague. Central to the oath was a pledge to always use fengshui to help and never to harm.

Use the methods and advice from this book to build a good home for yourself and your family. Protect yourself from harmful energy, but never direct harmful energy toward others. Land formations, environments, and exterior and interior architectural features are not only associated with positive or negative energies, they can actually affect our thoughts and emotions. Therefore, minimizing aggressive energy in your home will not only lead to harmony among those who live there but can also help you to be kinder and more considerate to others.

My hope is that with this book, more people will understand the need and want to make their home a healthy, harmonious, and prosperous place to live in. A helpful and cooperative neighborhood starts with good fengshui in individual homes. The more homes with good fengshui, the more harmonious the neighborhood. The less aggressive architectural and environmental features in a neighborhood, the less conflict in a community.

Use fengshui to help yourself, help others, and build a neighborhood that is a good community for you and everyone else to live in.

INDEX